Sustainability Thinking

Epaminondas Bellos

Sustainability Thinking

From Evolution to Constitution

Springer

Epaminondas Bellos
Pretoria, South Africa

ISBN 978-3-032-11243-9 ISBN 978-3-032-11244-6 (eBook)
https://doi.org/10.1007/978-3-032-11244-6

This Springer imprint is published by the registered company Springer Nature Switzerland AG
The registered company address is: Gewerbestrasse 11, 6330 Cham, Switzerland

In Gratitude to the Memory of My Parents

Contents

Chapter 1
Prologue and Other *Firsts*

It has been said that astronomy is a humbling and character-building experience…To me, it underscores our responsibility to deal more kindly with one another and to preserve and cherish the pale blue dot, the only home we have ever known—Carl Seagan

I am a *citizen-philosopher*—I embrace the incompleteness of knowledge and wisdom and the paradox that arises from knowing that not everything is knowable. I am a kindred spirit—if not a kindred mind—with the ancient Greek philosopher Aristotle (384–322 BC) who was a prolific natural philosopher whose insights were so powerful in their respective contexts that they first instigated breakout human progress and then due to their overwhelming dominance paradoxically stalled further development for over 1000 years until the advent of modern science. Aristotle developed empiricism and introduced the idea that universal truths can be arrived at via observation and induction, thereby laying the foundations of the scientific method. His writings profoundly influenced subsequent Islamic and European scholars and were only supplanted by the Copernican Revolution starting in 1543 [1]. Aristotle's science is one of the best examples of the power of *firsts*, which are understood here to be the path-setting events and conditions, which may or may not be subject to some form of subsequent human control. From the big bang to the emergence of life, *firsts* set paths, sometimes unpredictably or irrevocably so. Imagine unravelling the chemical reactions that led to life back to the primordial soup from which it is thought to have emerged. In the realm of human affairs, the ultimate root source of greater understanding is the study of human thought itself as the first *first*. As Aristotle himself pointed out in his *Posterior Analytics,* reasoning can never be completely autonomous because it always starts from premises and relies on inferential rules which reasoning itself cannot supply. Some other faculty must provide our sense that some starting points are more plausible, legitimate, or promising than others so as to anchor and enable reasoning. The human faculty of intuition is the mechanism that Aristotle proposed as how we obtain these first premises and first principles [2, p. 36]. Whether in science, mathematics, economics, politics, or philosophy the guiding influence of *firsts* (whether right or wrong) may be found in definitions, axioms, assumptions, biases, fallacies, hypotheses, theories, self-evident propositions, methodologies, and ideologies to name a few. Consequently, acknowledgement and careful consideration of relevant *firsts* is a must if correct paths are to be embarked upon. As a hopeful *first*, let us presuppose

E. Bellos, *Sustainability Thinking*, https://doi.org/10.1007/978-3-032-11244-6_1

that humanity is on a path to become increasingly wise within a context of complexity.

In 1988 as a Master's student in public administration at the University of New Mexico, I concluded in my unpublished professional paper, which was required for graduation, that intuition could be adopted as the mechanism for the wide-angle scanning envisioned in Amitai Etzioni's *mixed-scanning* model of decision-making, especially in ill-structured decision contexts. Etzioni's model aimed at compensating for the limitations of both the rationalist and incrementalist models of decision-making. To do so, the metaphor of a wide-angle weather observation camera that covers the broad sky, but not in great detail, together with a second weather camera that settles on areas shown by the first camera to require more in-depth examination captured the essence of this "third-way" of deciding [3]. Much later I discovered that mixed-scanning was in fact grounded in our biology, namely, the magnocellular vision pathway which generates wide-angle, low-resolution images for our brain, while the parvocellular vision pathway provides narrow, high-resolution images [4]. Mixed-scanning occurs not only in the visual centres of the brain, but also in its verbal ones. The left hemisphere narrows its attentional focus on highly related words in contrast to the right hemisphere which activates a broader range of words. The net result both visually and verbally is a right → left → right progression [5, pp. 40–46]. From biological substrates to decision-making metaphors, the most useful application of mixed-scanning for our purposes will be the dynamic pivot that it can facilitate between the individual and the collective engaged in decision-making and decision-taking.

Now fast forward to 2015, when as a seasoned Doctor of Philosophy in Public Administration from the University of Colorado at Denver, I encountered Daniel Kahneman's book *Thinking, Fast and Slow* as part of research I was conducting on energy transitions. Kahneman, a trained psychologist and Nobel Prize winner in Economics, posited that the human brain consists of two systems: one fast, one slow. These systems were described as one being intuitive, and the other rational. The similarities with my 1988 paper were intriguing, but not all that obvious. Both counterposed two types of thinking, both involved decision-making, and both pointed to effects on public policy. The fundamental difference between the professor's and the student's conclusions was that for Kahneman time was *internal* to decision-making, whereas for me, time was *external*. Kahneman was describing internal mental processes as being intuitive or rational (fast or slow, respectively), whereas I was addressing time as being the key variable as to whether intuitive decision-making would be triggered in less structured decision contexts with less time to decide, while relying more on rational decision-making when contexts became increasingly structured with more time to deliberate. Years later I discovered that my thesis could be supported by the structure of the human brain itself in that the right hemisphere is more adept in dealing with new situations and when new information has to be distinguished from old, whereas the left hemisphere is more efficient in routine situations when things are predictable [5, p. 40].

It also became evident that the collective perspective in decision-making was not being adequately represented in the literature. In consequence, this book advances the term *metasophia* as the most suitable candidate word in English to represent humanity's transcendental collective knowledge, imagination, and wisdom—the origin of all of our progress—as well as unintended consequences. The addition of the prefix "meta" to "sophia" denotes something of a higher or second-order kind. Similarly, *metacognition* implies multiple perspectives, a balance of viewpoints, context adaptability, and epistemic humility from within a person-centric framework [6]. These are important aspects of wisdom but are highly individual and process-dominant, hence a better-rounded definition of wisdom also requires an emphasis on collective decision and action outcomes, as well as an open sharing orientation, all of which support the addition of the term metasophia ("meta-wisdom"). "*Mêtis*", from the ancient Greek, is an allied term of metasophia which denotes intelligence, craft, skill, cunning, the subtlety of art, and the power of invention. In Greek mythology *Mêtis* was the goddess from whom Athena derived her own wisdom or *sophia*. Ancient Athenians understood that no physical force was mightier than the mind [7, p. 6]. In what follows, the difference between sophia and metasophia is simply that sophia and mêtis refer to the individual, while metasophia refers to the collective. We may therefore say that the goddess Athena had sophia while the Athenians had metasophia. Likewise, we might say that an individual decision-maker may have sophia, while a decision-making collective may have metasophia.

We can also use the mind's eye to see from the perspective of the State as James Scott so elaborates in his book *Seeing Like a State* [8]. The subject of vision abounds in language. The English terms "evident", "obvious", and "intuition" have roots in verbs for seeing [9, p. 238]. Insight—inner sight—depends on a holographic-like analysis of an input in its various aspects, in deriving relevant messages from irrelevant noise, and in identifying patterns in a bricolage until it becomes saturated with meaning [10, p. 94]. In science, we accomplish similar objectives by a number of sophisticated instruments designed to literally see what we cannot with the unaided eye ranging from the Atomic Nanoscope and Functional Magnetic Resonance Imaging, to Ground Penetrating Radar, and the Event Horizon Telescope, which in April 2019 produced the first direct image of a black hole in the centre of the super-giant elliptical Messier 87 galaxy [11]. In rapid succession, the James Webb Space Telescope was launched in December 2021, and after undergoing months of commissioning where its mirrors were aligned and its four instruments calibrated, its first images were publicly released in July 2022.[1] Among previously unseen regions of the universe, the telescope detected the first signs of water in the atmosphere of an exoplanet, opening the door to the study of the atmospheric composition of hundreds of other systems [12].

Although considerable reliance is also placed on the other human senses, they are less privileged than vision; they do not, for instance, enjoy the assistance of direct quantification by a universal constant—the *candela*—which is used to

[1] In partnership with the European Space Agency, Canadian Space Agency, and the Space Telescope Science Institute, available at https://www.nasa.gov/webbfirstimages

Table 1.1 Elements for collective wisdom and action

Thinking processes	Thought outcomes
Ethics	Sustainability
Empathy	Compassion
Intelligence	Foresight
Attention	Focus
Imagination	Innovation
Knowledge	Memory
Experience	Norms
Expertise development	Epistemic equilibria
Techne/episteme	Phronesis/sophia

measure the sensitivity of the human eye. Our everyday experience, however, is based on the more familiar 20/20 standard for eyesight correction. Humans are unique among primates in having white in their eyes which aids in mutual gaze-following, suggesting that human vision evolved to support cooperative social interactions [13, p. 222]. Mutual gaze-following also occurs in other social animals like dogs. My first dog, Philida, a sweet Rhodesian Ridgeback, "spoke" with her eyes by glancing rapidly and persistently at the window ledge where we kept her treats. Visual thinking is an earlier form of mental activity than thinking in verbal concepts—we dream mostly in pictures—not in words [10, p. 194]. The structure of the periodic table of elements came to Dmitri Mendeleyev in a dream [5, p. 107]. Using eye-tracking technology, at times measured in milliseconds, experiments have shown a clear connection between the way a participant's words are ordered, and the way that person sees a picture. Psycholinguists call this relational encoding and have found that when people are putting their thoughts into words, their mental processing may be ordered by the structure of their language [14]. Interestingly, the development of generative artificial intelligence has followed a similar trajectory as the evolution of the human brain in disassociating the processing of "pictures" and "words" by a three-stage process involving perceptual, lexical, and contextual representations [15]. It is only the human brain, however, that can recombine visual inputs with word associations to imbue overall salience to an experience in part by running these through the personal memory and emotional centres of the brain [16, pp. 64–65]. For present purposes consider the proposition that collective wisdom and action are made up of the ethics, vision, and qualia suggested by Table 1.1 which will be synergistically needed to attain sustainability.

Out of the Blue

The vast mystery of biology is that life should have emerged at all, that the order we see should have come to pass—Stuart Kauffman

The general type of inquiry we will follow is called "coarse graining" by scientists, which basically means that we may dispense with details that are not important or relevant to the main subject at hand. To elaborate: when a detail in a photograph is so small that it needs to be greatly enlarged to be clearly seen, the enlargement may show only a few dots conveying a rough image instead of a clear picture of the detail. The graininess of a photograph supplies a limit on the amount of information it can hold. If the film is very grainy, the best the image can do is to give a rough impression of what was photographed, that is, it will be exhibiting very coarse graining [17, pp. 29–30]. The same conditions hold true for normal digital photography. Images of all types are excellent vehicles of information and communication. Take, for example, the impact of the first image of Earth taken in colour from the moon by Apollo 8.[2] The colour film proved to be the missing ingredient; a similar photo had been taken two years earlier, but without the dazzling blue of the Earth. And apparently the Earth is becoming more blue as its oceans warm [18]. In the words of space shuttle Commander Jeff Ashby "circling the entire planet in 90 minutes, you see that thin blue arc of the atmosphere, seeing how fragile the little layer is in which all of humanity exists" [19, p. 129]. When the Earthrise photo was republished on the cover of *Life* magazine, billions of people could also look back at Earth in all its glory. That this life-giving planet was the same thing as Creation was a message the astronauts reinforced on the same day with their reading from the *Book of Genesis*. A great deal of earth science has since been inspired because of this image. NASA's spacecraft and their cameras have helped to locate ancient monuments covered by forest canopies, oil leaks in the oceans, the drying up of the Aral Sea, unusual weather across the globe, shrinking glaciers, and other signs of a planet that has changed a great deal since 1968 [20]. On the fifty-year anniversary of taking the photo, Apollo 8 photographer William Anders remarked that "we set out to explore the moon and instead discovered the Earth". In *Life* magazine's *100 Photographs that Changed the World*, wilderness photographer Galen Rowell called the Earthrise image "the most influential environmental photograph ever taken". Author John Wilford called its publication the beginning of the environmental movement [21]. At the close of 2022, NASA's Artemis I mission reminded us once again of NASA's collection of Blue Marble imagery and the stark visual difference—and all that it implies—between the blackness of space, the grey moon, and the blue Earth.

Now let us turn our attention to fine-grained analysis so as to draw a clear distinction between coarse graining and more fine-grained resolutions so as to be able to thoughtfully situate all subsequent discussion within the spectrum of possibilities between the two.[3] In our example this can be demonstrated down to the level of human neurons, in particular, those involved in vision and speech. The perception of colour in mental imagery activates only the right side of the human brain as suggested by studies that show the right hemisphere as being more attuned to colour discrimination and perception; however, the brain regions involved in *conscious identification* of colour seem to be left-sided perhaps because this involves

[2] Available at https://www.nasa.gov/multimedia/imagegallery/image_feature_1249.html

[3] In forecasting, the term *resolution* is used differently. It denotes the equivalent of the span of a "camera", while *calibration* is used to denote the degree of detail in a "photograph" [22, pp. 62–63].

categorisation and naming [5, p. 63]. Scientists have traced colour vocabulary all the way from the genes that prescribe colour perception to the final expression of perceived colour in language. Yet colour does not exist in nature. Visible light consists of continuously varying wavelength without intrinsic colour. Colour is imposed by photosensitive cone cells in the retina that transmit it via nerve cells to the human brain. The cone cells contain blue, green, or red photosensitive pigments which absorb the light energy coming into the eye. This energy is converted to electrical signals that are relayed to the retinal ganglion cells forming the optic nerve. At this stage the wavelength information is recombined into signals that are parsed along two axes: one on a continuum from green to red, the other from blue to yellow. How strong the signal is tells the brain, for example, how much blue or yellow the retina is receiving, yellow being a mixture of green and red. Collective information from vast amounts of cones and ganglion cells is delivered by a relay station near the centre of the brain to the primary visual cortex at the rear most part of the brain. The visual information, now colour-coded, is used by the brain together with other information such as memories. The brain thus converts the continuously varying wavelengths of visible light into an array of discrete units we call the colour spectrum [23, pp. 175–180].

The Munsell array is a human invention that depicts the colour spectrum from left to right, and in brightness from the bottom to the top, the latter to measure light intensity. Our genes prescribe that we see wavelengths of light in a particular way. Our further predilection to reduce the world into units and label them by means of words causes us to accrue up to eleven basic colour units in a specific order [23, pp. 178–179]. When subjects were asked as part of an experiment to match the principal colour terms of their respective languages (twenty of them) to the closest corresponding Munsell chips, the resulting clusters of terms coalesced around the principal colours of blue, green, yellow, and red. In a later study it was established that societies use from two to eleven basic colour terms. Languages with only two terms use them to differentiate between the colours black and white. Languages with seven terms have words for black, white, red, green, yellow, blue, and brown *in this order*. The observed hierarchy throughout the other six levels stops at the given configuration on the seventh level, with no further precedence given to the four additional colours of grey, purple, orange, and pink. In addition, colour terms at times acquire extra meaning to denote characteristics such as texture, indelibility, luminosity, and freshness. For example, in ancient Greek *chloros*, which is usually translated into English as simply "green", originally conveyed the moistness or freshness of green foliage as well [23, p. 179]. And this evening we would be well advised to remember that light at dusk is red yet we don't suddenly see leaves as reddish green—they still appear green to us because our higher-level neural machinery compensates for the less sophisticated colour areas of our "tell-tale" brain [16, p. 95]. Now also consider that the human eye has an actual blind spot where the optic nerve attaches to the retina, yet we never see a blank area in our visual field. Our perceptual system uses information from adjacent areas of a given visual field to predict what should be present in the blank area and then actively retouches the image to fill the blind spot [9, p. 11].

The analogy to vision is well suited for demonstrating the need for both scientific and non-scientific perspectives in differing combinations. The illustrations throughout this book are a testament to the practical value of images as an aid to the understanding of both ideas and words. Simple rules can also help, such as "no story without numbers, and no numbers without a story; or no model without facts, and no facts without a model" [24, p. 136]. We will return at times to the issues of seeing, visibility, and accessibility as a heuristic of this book.

Learning from Evolution, Looking to the Future

No man ever steps in the same river twice, for it's not the same river and he's not the same man—Heraclitus

Only by knowing the whole human story can we truly understand who we are, and how we came to be that way. From the first arrival of humans around six million years ago to join other already diversified animal species human thinking has been evolving. Yuval Harari in his popular book *Sapiens* has proposed that a random genetic mutation about 70,000 years ago led to a "Cognitive Revolution" marked primarily by the emergence of language. Iain McGilchrist maintains that the attainment of the skill of imitation may explain the otherwise inscrutable rapid expansion of the brain in early hominids since it would have energised a sudden take-off in the speed with which we could adapt and change ourselves in a direction of our own choosing [5, p. 253]. New evidence has increased speculation that perhaps both Neanderthals and Sapiens had the ability for abstract thought and language, thus the responsible genetic mutations(s) may have occurred earlier in the common ancestor of both species [25, pp. 174–175]. If this is the case, then the conditions for the start of the Cognitive Revolution may have to be pushed back from 70,000 to about two million years ago when humans spread from Africa to Eurasia in the midst of the evolution of different human species [26, pp. ix-x]. Accomplished neuroscientist, Joseph LeDoux, has proposed that somewhere in this time span emotions became the centre of gravity of the human brain and could not exist in the form we experience them today without our early hominid ancestors having evolved language, hierarchical relational reasoning, and noetic consciousness. These capacities made it possible for ancient survival circuits to be integrated into self-awareness and framed in terms of semantic, conceptual, and episodic memories. These could then be interpreted in terms of personalised self and emotional schemas to guide behaviour in the present and also to plan for future emotional experiences [25, p. 368].

In their sweeping survey of life on planet Earth, Maynard-Smith and Szathmary identified eight transformational transitions in the evolution of complexity of living things. Astonishingly, in each case the transition was marked by only two common fundamental processes: first, there emerged some new form of cooperation together with corresponding interdependence; second, each type of cooperation was made possible by an accompanying new form of communication. Early humans made reference to things, but their own relation to that referential material was left implicit (maybe by a facial expression or vocalisation), but not as an intentional part under the communicator's voluntary control in decision-making. In the further development of human language, the essential social context in which such communications occurred was collective decision-making as it occurred regularly in collaborative activities. These exchanges took the form of reason-giving discourse to convince others, especially in situations of low trust. Convincing others enhanced individual fitness so humans evolved human reasoning abilities not for getting at the truth, but rather for convincing others of their views. Directing these processes inwards internalised the communicative act to one's self as recipient, resulting in a kind of inner dialogue. This distinguished the "self" versus the "other", including holding the other to the normative criteria of intelligibility of the reference cultural group. The outcome was that the various conceptualisations and propositionally expressed thoughts in language became increasingly inferentially interconnected in a vast "web of beliefs" such that each element gained significance from its inferential relations with others. The resulting web of beliefs and humans' ability to navigate this web became foundational for the ability to engage in individual reasoning [27, pp. 32, 102, 110]. As modern humans began engaging in *cooperative argumentation* to convince each other, implicitly accepted linguistic norms emerged. These norms operated such that individuals who contradicted themselves from one assertion to the next, or changed the meaning of their terms mid-stream, or held something to be both true and false without explanation, were basically ignored or excluded from the group decision-making process. The process of cooperative argumentation then was the special language game within which human norms of rationality came to govern all those who wanted a voice in collective decisions, such as in epistemic, judicial, and political decisions [27, pp. 117–118].

In his foundational book in anthropology, *A Natural History of Human Thinking*, Michael Tomasello sets out the cognitive schema of joint collaborative activity comprised of individuality plus simultaneous sharing that occurs at a dual level of social engagement which explains and lays the foundation for almost everything that is uniquely human [27, pp. 48–49]. It represents the joint intentionality architecture of quintessentially human forms of cooperative communication involving intentions and inferences about different perspectives which ultimately provide the foundation for cultural conventions, institutions, and norms. This fundamental dual-level cognitive schema of concurrent jointness and individuality has been scaled up by modern humans into a group-minded cognitive model of individuality and objectivity. Human group-mindedness marks a profound shift in the ways of both knowing and doing thereby "collectivising" human joint intentionality. Everything is made generic to fit anyone in the group in an agent-neutral manner which results in a kind of collective perspective, experienced as a sense of "objectivity". This makes possible the *objective-reflective-normative* thinking that modern human thought has

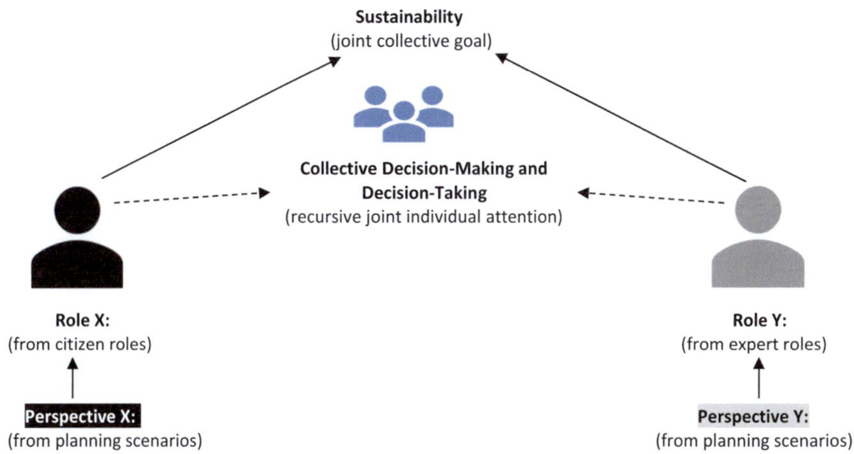

Fig. 1.1 Sustainability as dual-level joint collaborative decision-making and decision-taking. (Adapted from Michael Tomasello [27])

become from its *shared intentionality* and *dual-level joint collaborative decision-making* sources[4] as shown in Fig. 1.1.

Michael Tomasello proposes that as early human populations grew they tended to split into smaller groupings leading to recognisable tribal organisation in which a number of different social groupings were still a single supergroup or protoculture. This meant that recognising others from one's own cultural group and the recognition by others of ourselves became critical. Such two-way recognition was far from trivial because only members of a cultural group could be counted on to share in-group skills and values so as to be knowledgeable and trustworthy collaborative partners. Teaching others to do things, especially one's children, became a good way to assist their functioning in the group and to ensure even more conformity along the way. Teaching and conformity then led to cumulative cultural evolution characterised by the "ratchet effect" in which changes of a cultural practice stayed in the population with some fidelity until some individual invented some new and improved technique, which was then taught and conformed to until an even better technique ratcheted things up again. Tomasello argues that great ape societies do not display the ratchet effect or cumulative cultural evolution because their social learning is fundamentally exploitative and not cooperatively organised in the human way via teaching and conformity, which constitute the ratchet that prevents individuals from slipping backwards. The new sense of group identity characteristic of modern humans was in consequence expanded, not just in space to in-group strangers, but

[4] These are unavoidably characterised as hypotheses by Michael Tomasello since it is impossible to historically reconstruct past human thinking via hard evidence. Shared intentionality is comprised of individual, joint, and collective intentionality.

also in time to descendants and ancestors of the group—“this way is the way *we* have always done things—it is a part of who *we* are”.

Modern humans participate not only in joint intentionality with fellow individuals, but also in collective intentionality with the entire cultural assemblage [27, pp. 4–6,48–49, 93,113,140–141]. As cultural practices were cooperatively handed down across generations, adults altruistically teaching and youngsters conforming, the resulting cumulative effect was that the *we* became an enduring culture to which we (past, present, and future) are all committed. Human populations thus became more than a loosely structured pool of collaborators; they became self-identified cultures with their own memories and histories. The overall process was generally the same: a change in ecology led to some new forms of collaboration which necessitated for their coordination some new forms of cooperative communication, and then together these created the potential that during ontogeny, individuals could construct through their social interactions some new forms of mental representation, inference, and self-monitoring for use in their cognition. The first clear signs of distinct human cultures appear with modern humans beginning at the earliest some 200,000 years ago [27, pp. 31, 83–84]. Humans have created genuine evolutionary advances via new forms of cooperation supported and extended by new forms of communication. This has led to new forms of cognitive representation, self-monitoring, and inference amounting to new kinds of thinking. Culture and language provided an agent-neutral context within which a new form of sociality led to a new form of human thinking, namely, *objective-reflective-normative thinking*. Modern human’s creation of collective intentionality from its earlier subsidiary forms eventually led to a kind of agent-neutral, “objective” thinking, inclusive of cultural and linguistic representations; processes of inferring that were reasoned, reflective, and aimed at the truth; and normative self-governance in which individuals monitored and adjusted their thinking to fit with that of their group [27, p. 141].

Human culture is an aggregate case of emergence that arises in human groupings due to our social systems and higher-order thinking which allow for the cumulative retention and exchange of ideas. This means that the influence that culture has been exerting on our evolution has been increasing over the past ten to forty thousand years [28, p. 388]. A key insight of evolutionary thinking is that—in contrast to the metaphor of the invisible hand—the pursuit of lower-level parochial interests, such as short-term individual, corporate, or national interests, is far more likely to undermine than contribute to the global public good. Human cooperation requires identifying the most relevant group, establishing norms that define the welfare of that group, and establishing mechanisms that reward good behaviour and supress bad behaviour [29]. We will pick-up the human evolutionary story at the dual level of collaboration when full human cooperation emerged and traverse our chosen path, ratchet-like, from “evolution to constitution” suggestive of a distant attractor state that we might think of as sustainability driven by “self-help democracy”. The purpose of focusing on sustainability is to hold it out as the primary glue that ought to hold society together. Shared meaning undergirds shared intentionality which undergirds a shared goal—these are the fundamental elements necessary for collective thought metabolism and thoughtful action in the proposed schema. To this end, Fig. 1.1 will serve as our metaphorical map by travelling from the goal of sustainability backwards to its proposed modern roots derived from democratic perspectives generated by planning scenarios and refined by citizen and expert roles. Much of the more recent history of literate civilisation may be regarded as an evolution of ever more powerful tools for thinking: libraries and the internet to hold memory, universities and labs to create new knowledge, artists' and architects' studios to expand creativity, hospitals and doctors to heal, and firms and markets to coordinate production. Beyond Google's ambitious goal of copying the entire World Wide Web to represent it and then search it [24], research that explores the ways and means of global wisdom aggregation is a natural conclusion if we are to move steadfastly towards a sustainable future.

What This Book Is About

Ideas are the source of all things. Ignorance is the root cause of all difficulties—Plato

What follows is at the intersection of environmental humanities and political philosophy; democratic governance and decision theory; and at the intersection of biology and psychology. The book's storyline emerges from the origins of individual and collective biological evolution to better understand the path to be taken towards national constitutions that draw on sustainability thinking for sustainable action. *Sustainability Thinking* concerns the resilience of democratic governance and the capacity of our environment to sustain a rich variety of life—it is a "why-to" book. Its objective is to promote better democratic decision-making and better expert decision-taking. This should serve to derive the most benefit from citizen diversity while drawing on the combined knowledge and experience of subject matter experts. We will engage in *perspective taking* on sustainability as a global risk; being reliant on sound collective decisions, ways and means for *wisdom-making* will be undertaken; staying mindful of the "Greek tragedy of the commons", democracy will be recruited as the overarching mechanism for *attention-taking*; assuming diversity to be more equitable in *perspective-making*, four initiating sustainability scenarios for the future will be set out; and to promote thoughtful practices the need for collective *role-taking* by both citizens and experts will be highlighted. Most importantly, a renewed reverence for all life will be invoked to serve as the central societal value needed for sustainability.

Sustainability Thinking is also a "how-to" book—it is a manifesto for individual citizens to become more active and informed members of society. The principal message and argument put forward is that all existing and forthcoming sustainability solutions ultimately depend on various forms of decision-making and decision-taking—yet as the overarching keys to success—remain under addressed as subjects in the literature and solution space. Two crucial parameters—time and agency—are currently underrepresented or under emphasised. This book seeks to inform the goal-driven, risk aware, unencumbered aggregation of thought and knowledge required, as well as to instil the sense of urgency necessary in those entrusted to act on the behalf of each and all levels of humanity. Stated succinctly, our approach will be to prime metasophia and inform decision-making for better sustainability outcomes. To do this, we will need to contribute to the reasoned consumption of information, ideation, as well as critical and open-minded thinking as fundamental intermediaries to collective wisdom.

The outlook of *Sustainability Thinking* is optimistic *and* realistic guided by Al Gore's refrain that "we must not leap from denial to despair. We can do something and we must". To close the identified gaps, this book will contribute to the stock of possible solutions by: (i) highlighting the scientific basis of individual and group decision-making and decision-taking; (ii) recommending a cluster of citizen and experts' actions to aid sustainability; (iii) proposing a system of "self-help" democracy and the democratisation of philosophy; and (iv) introducing a unified theory of collective mind and collective action.

In doing the research for this book the broader lesson I learned is that a practical philosophy is needed because unlike "knowledge management" there is no academic discipline that addresses "wisdom management". There is no Nobel Prize for wisdom. And in the age of artificial intelligence (AI), can we speak of wise algorithms? We already speak of artificial intelligence, but not of artificial wisdom. Put differently: we should live up to our current designated species name—*Homo sapiens sapiens*—man the wise.

Chapter 2
Complexity as Context

X counts as Y in context Z—John Searle

In 1923, C. Lloyd Morgan proposed that an "emergent evolution" could explain the world's increasing complexity, including the evolution of mind. He noticed that many of the most profound changes in living things have been mostly discontinuous with past evolution. The gradual process of natural selection was not singular but also involved jumps in complexity in a sort of qualitative punctuated equilibrium. The complexification of human cultures, especially human languages, accelerated cultural evolution making it faster than biological evolution. Combined with a recent understanding of the human impact on the biosphere, the concepts of sustainability and that of coevolution may lead to the possible harmonisation of cultural and biological evolution [30]. According to physicist and Nobel Prize winner, Murray Gell-Mann "one of the greatest challenges of contemporary science is to trace the mix of simplicity and complexity, regularity and randomness, order and disorder up the ladder from elementary particle physics and cosmology to the realm of complex adaptive systems" [17]. Almost all the laws of physics can fall under the rubric of simplicity, mainly because they can be expressed in an economical way by way of just a few compact mathematical equations, such as Einstein's theory of relativity, Newton's laws, and Maxwell's equations, all of which can be elegantly derived from the *principle of least action*, which is claimed to be loosely related to energy. This principle has contributed enormously to our understanding of the world around us, including our modern technological society. Of special interest to us is the Navier-Stokes equation that sought to describe fluid dynamics, and in particular, turbulence, which remains an unsolved problem in physics. Complexity theory and complexity science owe their development on it, and it continues to underlie the design of hydroelectric power stations, as well as inform climate change, and the hydrology of rivers and water supply systems, among other applications [31, pp. 71–72, 160].

One can legitimately wonder, however, whether some attempts at reductionism are not overly Procrustean, attempting to fit the natural and social worlds into perspectives and methodologies like the traveller in Greek mythology who was stretched or had his legs cut off so as to exactly fit the inn-keeper's bed. Consider the Game of Life and the human brain. Both consist of spatially connected simple

E. Bellos, *Sustainability Thinking*, https://doi.org/10.1007/978-3-032-11244-6_2

parts that rely on threshold-based rules. Both can produce any class of outcome. The Game of Life is played on a checkerboard using two rules: the "life rule" causes a dead cell with three live neighbours to become alive; the "death rule" causes a live cell with fewer than two or more than three live neighbours to die. These two simple rules can produce elaborate patterns and outcomes which can be thought of as calculations, not unlike those undertaken by the human brain. Albeit an artificial demonstration, the Game of Life shows us that even a few basic laws can produce highly complex characteristics, even a universal computer or human intelligence when acted upon by evolutionary forces [32, pp. 177–179]. But treating an organism like a simple machine is a type of reduction, approximation, or simplification that is precisely like a Procrustean bed. Evolution thrives on stressors, uncertainty, disorder, and randomness by taking advantage of shocks that enhance the collective fitness of the gene pool, while individual organisms within it remain comparatively fragile. Nature is content to let the game play out at the informational level, the genetic code. The code is not really concerned about the welfare of the organism itself. Clearly, there is an inherent tension between nature as a collective and the individual organisms of which it is composed [33, pp. 67, 81, 202].

In biology, the "fitness landscape" model has provided a powerful heuristic to illustrate the role of competing schemata. It assumes that species have variable traits that contribute to their fitness, usually understood to be their reproductive potential. If we plot a value for a trait on the horizontal axis of a graph and fitness on its vertical axis, we can produce a chart known as a fitness landscape in which points of high elevation correspond to high fitness. Plotting the values of a trait on the X-axis versus its associated fitness values on the Y-axis, we can move right on the X-axis and up the resultant "mountain" to its highest point and thereby obtain the optimal amount of a given trait for its corresponding fitness value. The landscape may become *rugged* when there are interdependent effects among more than one trait or variable, conditions which promote the likelihood of suboptimal outcomes shown as local peaks ("hills") on the landscape [32, pp. 327–331]. Although the use of fitness landscapes has been a useful concept in biology since its creation by Sewall Wright in 1932, it has advanced haltingly in the intervening years. For example, the question of how developmental mechanisms themselves change during evolution and how evolution then acts on these changes remains insufficiently explored. Consider that the environment acts on an entity, but the entity also acts on the environment, and both are dependent on feedback. George McGhee has proposed that the study of comparative morphology and the development of theoretical morphospaces can further the analytical usefulness of adaptive landscapes [34, p. 183]. According to John Stewart, genetic systems function similarly to nervous systems in that they are structured and organised by evolution to enhance their capability to discover more effective adaptations [35, p. 418]. Without direction *genetic drift* would push evolution squarely into the realm of chaos. Arthur Koestler describes such a trial-and-error process as being at a deeper level than natural selection:

> ...before a new mutation has a chance to be submitted to the Darwinian tests of survival in the external environment, it must have passed the tests of *internal selection* for its physical, chemical and biological fitness. The concept of internal selection, of a hierarchy of controls

> which eliminate the consequences of harmful gene-mutations and co-ordinates the effects of useful mutations, is the missing link in orthodox theory between the "atoms" of heredity and the living stream of evolution. Without that link, neither of them makes sense [10, p. 132].

Fitness landscapes have been used less successfully in other domains. In knowledge management they have been applied without the concept of fitness having been adequately redefined and are furthermore guided by the assumption that learning will always find the highest peak. Even with paradigm shifts and scientific revolutions, which are equivalent to a peak collapse, the multiple interlinkages of knowledge networks do not suggest the establishment of one smooth global peak (Fujiyama model), but instead a very rugged landscape which is likely to result in *path dependence*. Early choices determine future possibilities in path dependence highlighting the need for sound early decisions if complex systems are to be properly managed [36]. In human decision-making, cognitive dissonance is mostly handled by adjusting conflicting statements of fact, norms, or causal inferences so as to restore harmony in thought processes. Cognitive rigidities generally serve to deflect or avoid the forces of change, and in the process, contribute to the effects of path dependence [37, p. 77]. Human beings have agency; however, path dependence may be mitigated or converted into path creation [38]. The exploratory drive has been recognised to be a basic biological instinct which at times is even stronger than that of hunger or sex. Experiments have shown that curiosity and thrill seeking are an instinctual drive in birds, dolphins, rats, chimpanzees, and humans [10, p. 153]. But to add a human societal dimension, more nuanced explanations are needed.

Path dependence in economics has imported the idea of the "butterfly effect" from chaos theory which holds that small initial perturbations can have an outsized system effect, but this notion has been conflated with the theory that there are several perfectly stable alternative states, one of which will arise based on the unique initial conditions under consideration. The potential for endless "disequilibrium" essential to chaos theory is consequently missing from the economic analysis of path dependence. So, it remains an important question whether theories that overlook the sequence of events leading to equilibrium are ignoring something fundamental. One compensatory approach has been to propose that available paths come in three degrees. The paths taken under first- and second-degree path dependence cannot be improved upon given the state of knowledge and available alternatives. By way of contrast, third-degree path dependence involves the practicality of improvements in the path taken. Accordingly, for any allocation decision that might be predisposed to path dependence, we can relax the usual presupposition that individual choices can correct market failures so as to lead us to an optimal outcome [39]. A prominent example of economic path dependence is the world energy system, which despite the challenge of climate change would require a significant transition period to be converted from fossil fuels to cleaner forms of energy.

Diversity is generally espoused as being beneficial in many physical and social settings because it offers more choices and protection from unforeseen events. It follows that the corollary of this proposition is that simplicity renders systems fragile or vulnerable if a few controlling rules can be isolated and manipulated wilfully or wantonly. Thus, in safe, hidden contexts simplicity suffices. Conversely, in volatile exposed contexts complexity becomes a protective cloak. Take the example of DNA versus blockchain. Viruses are able to highjack DNA due to its relative simplicity in contradistinction to blockchain technology which relies on layered complexity to achieve robustness. It may be that DNA and blockchain are two credible examples of effective solutions to diametrically opposed contexts (simplicity versus complexity) and the types of threat that each may withstand. But the foregoing contention does not take into account *potential complexity*. A system may be fairly simple but capable of evolving with high probability into something much more complex given sufficient time. This is the story of *homo sapiens* who were not much more complex than the great apes but possessed the requisite potentialities to develop cultures of enormous complexity [17, p. 370]. Complexity as a subject is a rich, multifaceted concept that can be considered as a property with multiple attributes, or it can be considered as a measure [40, p. 26]. Simple binary formulations, however, do not suffice. The more nuanced "edge of chaos" expression captures the essence of all adaptive systems as they evolve or hover around the boundary between stable order and unstable chaos. It is at the edge of chaos where productive activity can be undertaken because the conditions in simple states are too static, while overly active ones are too chaotic [41, p. 129]. Life's dynamical persistence is driven by chemical disequilibria in local environments, as well as the larger disequilibrium between the solar photons of the universe compared to their thermal background [42]. Our version of "edge of chaos" conditions occurs at the confluence of open democracy and epistemic democracy as shown in Fig. 5.2. The entire "science of the social" is cloaked with yet more complexity given that any sought for objectivity is complexified by assumptions that function as a kind of observer, but this "observer" overwhelmingly affects what it is observing, and likewise what is being observed affects the observer [43, p. 80].

Complex Adaptive Systems

> *An event that would be unthinkable in a hundred years may be inevitable in a hundred million*—Carl Sagan

Complex adaptive systems bridge complex systems theory and evolution principles and encompass a wide range of physical and social phenomena [35, p. 445] from the behaviour of organisms in ecological systems to the evolution of human societies. The etymology of the English word "complex" has both Latin and Greek origins. The Latin *complexus* leads to "complex". The mathematical term "symplectic" is derived from the Greek πλεκτός (plektos), literally meaning braided together [44]. Adaptive means dynamic change to better fit a changing environment. Galaxies are complex, but they are not adaptive. An alphabet is simple, but not adaptive. Viruses may be the simplest organisms on earth, but they are highly adaptive. By way of contrast, the earth, human societies, and even individual human beings may be considered to be complex adaptive systems whether they occur individually or in nested configurations. In some complex adaptive systems, adaptations are a consequence of individual agents changing their behaviour in anticipation of future outcomes and positions of other agents. These tend to be non-linear relationships and the state of equilibrium only temporary, multiple, and moving [45]. As agents gain more strategic ability, elaborate dances of strategies may arise, with good and bad epochs, cycles, and crashes [36, p. 50].

Complexity science research suggests that there are two principal ways of dealing with the complexity of systems: *complexity reduction* and *complexity absorption*. The former attempts to reduce the complexity of a system and its interdependencies by the compartmentalisation of information in new theories, models, or rules thereby reducing the amount of information that needs to be processed by decision-makers. In addition, standardisation and modularisation of relevant techniques and technologies may be deployed. Conversely, complexity absorption may involve increasing the diversity of participants as well organisations, including the creation of new institutions. Decentralised decision-making, networking, and participatory methods are additional strategies to absorb complexity. Both of these differing approaches close the complexity gap between the system and its governance structures [46]. Of most interest for present purposes is the evolving complexity of decision-making. This complexity gap needs to be managed in light of its deleterious growth. Across the world, a more participatory style of decision-making is becoming predominant but one which is still in competition with authoritarian and technocratic styles. The environment, equity, human rights, gender, and even animal rights are gaining purchase and are rising to the level of actual decision criteria in public decision-making. New social actors such as non-governmental organisations, transnational companies, and citizen movements contribute to an increase in the number of dimensions used to define issues, problems, and solutions thereby giving rise to higher complexity [47, p. 364]. Although diversity of thought and background has the potential to help groups think more broadly and analyse information more deeply, that potential is realised in some contexts but not others. New research shows that people are more likely to promote inclusion and diversity when the message is more nuanced. Acknowledging complexity does not make communication less convincing, it makes it more credible; it does not lose the audience, it maintains its engagement while stoking curiosity [19, p. 174].

A complex adaptive system framework applied to the study of social-ecological systems (SES) can inhibit us from seeing humans as positioned outside or above nature, which is only valued in terms of its usefulness to social systems. Instead, humans can be seen to be fundamentally embedded in natural systems and to profoundly affect the Earth's biosphere in its entirety. A fundamental understanding of SES as complex adaptive systems allows us to engage and develop theoretical and methodological orientations that are responsive to the complex patterns and dynamics of real-world SES. Examples of emergent properties in real-world systems include compromised immunity in human biological systems; food insecurity in agricultural and economic systems; and poverty in social and economic systems. Social-ecological systems are intertwined through material and nonmaterial, or functional processes in which the same constellation can enhance stability while other interactions are working towards transformation into a different constellation. Complex adaptive SES are transitory phenomena and hence the process of identifying any constellation is always the product of being embedded in the complex web of relations that link knower and known [48]. This web of relations includes the power of democracies to adapt in the face of uncertainty by, among other things, recognising the need to create a centralised and unified clearinghouse where they can process the world's complexity as a whole and keep accountable providers of inputs to whom certain powers may have been granted temporarily [49, p. 194]. Connecting human behaviour to context through a complex adaptive systems lens is critical to inform environmental governance and management for sustainability [50]. Here is Gell-Mann:

> if humanity does equip itself somehow with a measure of collective foresight—some degree of understanding of the branching histories of the future—a highly adaptive change will have taken place, but not yet a gateway event. The accomplishment of the interlinked transitions to greater sustainability, however, would be such an event. In particular, the ideological transition implies a major step for humanity toward planetary consciousness, perhaps with the aid of wisely managed technical advances now only dimly foreseeable. After the transitions, humanity as a whole—together with the other organisms inhabiting the planet—would function much more than now, as a composite, richly diverse complex adaptive system [17, pp. 19, 366].

Panarchy as set out in the book by the same name (named after the Greek god Pan) is an integrative concept of economic, ecological, and institutional systems that seeks to explain unpredictable change across hierarchies and scales. Sustainability has been described as the capacity to create, test, and maintain adaptive capability. Development, however, is the process of creating, testing, and maintaining opportunity [51, p. 76]. A nested hierarchy of adaptive cycles makes up the panarchy. The panarchy may consist of both natural and socio-economic subsystems. A valuable contribution of the adaptive cycle model is the *recursive attention* it brings on processes of capital release and subsequent reorganisation, which have traditionally been neglected in favour of emphasising growth and conservation. As depicted in Fig. 2.1, the first fore-loop (from r to K) shows a slow, incremental phase of growth and accumulation. The second back-loop (from Omega to Alpha) represents a rapid phase of reorganisation leading to renewal. Adaptive cycles help to explain how complex adaptive systems can, temporarily, generate novel recombinations that are tested during longer periods of capital accumulation and storage. These windows of opportunity open briefly but without triggering cascading instabilities of the whole panarchy due to the shock dissipating architecture of nested hierarchies. In essence, the larger and slower segments of the hierarchy provide the memory of the past and of the distant that allows for the recovery of the smaller and faster adaptive cycles [52]. It may be, however, that the neglected "X" offshoot at bottom left of Fig. 2.1 is the most consequential because it implies going *through* resilience and onto transformation. Transformation requires accidental or deliberate "outside the box" trajectories in contrast to resilience which seeks the maintenance of an existing system's identity, feedbacks, structure, and functions [53]. Transformational energy transitions of the past are representative examples, and a transition to a low-carbon global economy is a possibility (perhaps probability) of the future. A missing link in the panarchy framework, which this book tries to address, is an explicit accounting of the role and effects of joint decision-making by

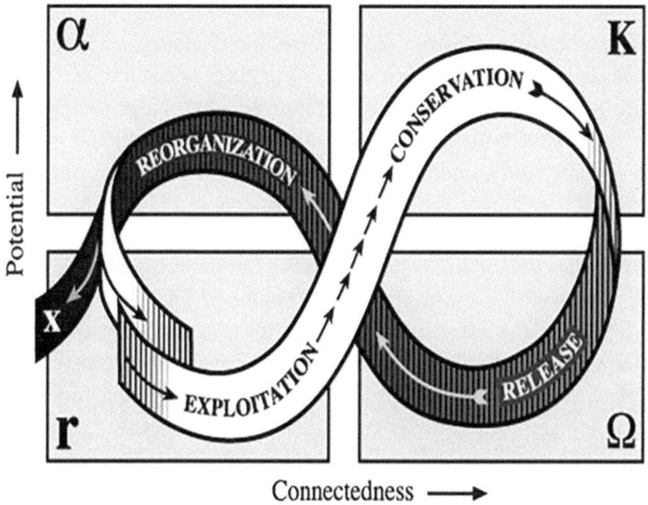

Fig. 2.1 Phases of the adaptive cycle in complex adaptive systems. (Adapted from [55])

policymakers and citizens. To some extent Elinor Ostrom's concept of collective choice rules within her general framework for analysing the sustainability of social-ecological systems covers this issue [54], but neither of these "system of systems" approaches rises to the needed level of treating the Earth as one whole, complex, integrated, *adaptive system.*

Complex adaptive systems (which is what Earth, humans, and humanity itself are) can respond differently depending on the environment. In a complex context, rigid rules are likely to fail because they place too much emphasis in the current state of knowledge, whereas systems that allow for revision due to low probability but high impact events may be suboptimal in the short run yet prove to be wiser in the long run. This is a major lesson of adaptive management research [56]. We will persist as individuals only if we manage to persist as *Homo sapiens sapiens*. We do not have enough time for biological evolution to come to the rescue—it is too slow of a process. We have to rely on the more rapid avenues of change—cognitive and cultural revolution, which in turn depend on our autonoetic brains and the individual and collective choices that we make. Indeed, in the end, it is consciousness in which we must place our trust [25, p. 379].

The Power and Complexity of Language

Scientists should be vigilant…and especially should guard against tacitly granting reality to things simply because we have words for them—Francis Bacon

Human language is central to our being and who we are as a species. The evolutionary developments of language have been codified by ontogeny through to this day but still remain substantively mysterious since we do not really understand how human infants acquire their native languages. In any human language, there are an unlimited number of possible sentences, which can contain a multitude of variable clauses, each with modifying words and phrases. Sentence length is limited only by the time available and by the patience and memory capacity of the speaker, listener, and writer. What is more, there is typically a vocabulary of many thousands of words to work with. In this context there is no chance that a child will hear or try to speak every possible sentence and enter it in a look-up table. Yet at the end of the real language learning process, a child can tell whether a previously unheard sentence is grammatical or not [17, p. 52]. Beyond the wonder of a child acquiring language, human language universals include an inverse ratio between the length of words and the frequency of their use, deployment of grammar, phonemes, metonymy, polysemy, and antonyms [28, p. 10]. And the sounds of words are evocative rather than arbitrary, suggestive of the experience that things refer to. Those with absolutely no knowledge of a particular language can nonetheless correctly guess which word matches an object in what has become known at the "kiki/bouba" effect; "kiki" suggesting a spiky-shaped object, in contrast to "bouba" suggesting a softly rounded object [5, p. 119]. And philosophers call the verb "intuiting" a success term, while linguists use "factive" to describe the intermediate state between hidden fact and the laws of nature [57, pp. 29–30].

The word "time" is the most frequent noun in the English language, and though ubiquitous, we cannot see, touch, or smell time. One factor that influences the perceived direction of time is writing direction. People who read native text arranged from left to right (English or Spanish, for example) tend to lay out time as proceeding from left to right, and people who read text arranged from right to left (Arabic or Hebrew, for example) arrange time from right to left. Patterns in spatial-temporal metaphors are also involved in perceived time reversals. For instance, English and Spanish spatial metaphors put the past behind the observer (the worst is already behind us) and the future in front (the best is still ahead of us). In Aymara, this pattern is reversed, and the future is said to be behind the observer while the past is in front. Beyond event ordering, interesting cross-linguistic differences have additionally been noted in a separate aspect of time—representations of duration. English speakers talk about duration more often in terms of linear distance (say, a long time), while Greek speakers talk about duration more often in terms of amount (say, *poli ora*, "much time"). Research shows that people who talk about time differently also think about it differently, and language not only reflects the structure of our non-linguistic mental representations, it can also shape those representations in fundamental ways that can be observed even in low-level perceptuo-motor tasks. People in different cultures have been shown to differ in whether they think of time as stationary or moving, limited or open-ended, as distance or quantity, horizontal or vertical, oriented from left to right, right to left, front to back, back to front, or in cardinal space (say, East to West). The metaphors we use to talk about time have both short-term and long-term effects on how we conceptualise and reason about this fundamental domain of experience [58]. At the whole language level, some languages like Maya are more long lived than the civilisations they served. Even when they seemingly disappear, some languages have lingering influence considering, for instance, that more than half of English words have Latin or Greek-via-Latin origins [59, pp. 479–480]. Human language developed in its current form to enable us to refer to whatever is not present in experience: language helps in the temporal "re-presentation" of experience making it useful for communication and thinking for some purposes while reducing it for others. The denotative elements that enabled precision of reference and planning have become a left hemisphere specialty, while the connotative and emotive elements have become a right brain specialty where all inputs are recombined to make sense of their context, including tone, irony, humour, and metaphor. Language hugely expands the range of reference of thought and our capacity for planning since we can't easily direct others to carry out our intentions at a distance [5, pp. 114, 125]. And language asks questions, facilitates imitation, and enables metaphor.

The Communication and Miscommunication of Science

Do not act on what you've heard many times; or on tradition, rumor, scripture, surmise, axiom, specious reasoning, bias, or deference to monks—The Buddha

Any complex idea is difficult to communicate; however, scientific concepts pose special challenges. New discoveries and research findings may require people to discard long-held beliefs or familiar frames of understanding of how the world works. And scientists tend to be so steeped in their own literature that they often fail to appreciate how inscrutable their findings might be to the general public [60]. Access to science needs to be understood as nuanced and multifaceted. People must be able to access scientific information, translated and actionable by a non-specialist audience [61]. Instead, much is "lost in translation" from the time of scientific discovery to the literary world and popular culture. Certain terms interpreted vaguely or inaccurately are often the main survivors of the journey from the technical publication to the popular magazine, paperback, or website. Important qualifications, and sometimes the actual ideas themselves, tend to get lost along the way. One can make the case that words like "chaos" and "energy" antedate their use as technical terms, but it is in their technical meaning that confusion is introduced, not in the original meaning of the words [17, p. 27].

Matters get even worse when scientists defend science too broadly pointing out that its self-correcting nature is not fully appreciated by journalists and the public. However, their inherent conflict of interest is not usually made evident in such arguments thereby undermining science's overall value over time because while any self-correction is underway real hazards or societal damage may materialise due to their initial missteps. What is more, any scientific missteps will vary both in magnitude and effect making holistic identification, evaluation, and society-wide correction all the more difficult. For example, the Sugar Research Foundation in the United States sponsored and published through a prestigious medical journal its first coronary heart disease research project in 1965, which singled out fat and cholesterol as the main dietary causes of heart disease and downplayed evidence that sugar consumption was also a significant risk factor. In marked contrast, a 2016 study with no disclosed conflicts of interest found, together with other recent analyses of internal sugar industry documents, that the sugar industry sponsored a research programme in the 1960s and 1970s that successfully cast doubt about the hazards of dietary sugar while pointing to fat as the main dietary culprit in heart disease. The 2016 study's authors concluded that policymaking committees should consider giving less emphasis to food industry funded research [62], which is an obvious and timid recommendation.

Clearly, there is a special concern regarding the delivery of scientific information to policymakers and people who can use it directly in their lives. In a survey of about 400 science policy specialists by *Nature*, eighty percent of respondents thought that policymakers lacked sufficient understanding of science while seventy-three percent said that researchers did not understand how public policy formulation works. It is no surprise then, that seventy percent of respondents thought that governments are not routinely using their advice [63]. People at risk from certain threats, such as smoking and obesity-related conditions, could be helped by a better understanding of relevant science. Typically, this would involve substantial behavioural changes that could improve outcomes for such groups. The use of prescriptive scientific narratives has been proposed to help bridge the gap between the jargon of scientific journals and the interpretations of journalists so as to better convey scientific findings in a comprehensible manner to the public. This may help the audience to understand and more fully remember complex information thus enabling better personal decisions. By using such narrative formats, communication of scientific information may help its audience to reconsider long-held beliefs that may be inaccurate or incomplete and to develop a deeper understanding that can serve as a new framework for emergent information [60].

More broadly, the opposite seems to be happening. For example, truth is undermined by the now questionable mantras of objectivity in the US journalism, such as those relating to "equal time" and "telling both sides of a story" even regarding factual matters. While this may have been a reasonable or even laudable goal of fairness when it came to opinion-based issues, it has proved to be a disaster for science coverage. By allowing "equal time", the media has only succeeded in creating a *false equivalence* between two opposing sides of a subject when in reality only one side may have proved itself to be credible. This phenomenon has fallen comfortably into the lap of ideologues and their allies who have come to routinely embrace a self-serving high standard of doubt towards the facts they wish to discount in line with their pre-existing beliefs. This is not the abandonment of facts as such, but the corruption of the process by which they are credibly collected and reliably used to shape one's beliefs about reality. This rejection that some things are true irrespective of how we feel about them runs counter to the public interest, as well as the necessary respect for truth and the methods of inquiry that lead us there

[64, pp. 11,77]. To overcome binary bias, a good approach is to acknowledge the scope of perspectives across the range of an issue. For example, polls suggest that on climate change there are at least six categories of thought. Believers make up more than half of Americans, but some are merely concerned, while others are alarmed. The nonbelievers actually range from cautious to disengaged to doubtful, while some are dismissive. We might be fooling ourselves in thinking that only two opposing sides can represent complex issues. People are actually more inclined to think again if we present complex topics through the many possible lenses of a prism [19, pp. 165, 169], hence the inherent value of different planning scenarios as we will see in Chap. 9. It has been shown experimentally that topic and technique rebuttal are effective strategies (singly or in combination) to combat science denialism. Of these, technique rebuttal, such as pointing out that originators have set impossible standards that cannot be met, converts planted doubt into more realistic possibilities [65]. To assist, the World Health Organization has developed best practice guidelines to counter vaccine denialism which can be incorporated into technique rebuttal training in other domains [66].

And technology is not a panacea. Technology provides a different version and amplification of the Matthew Effect in science communication: "for unto everyone that hath shall be given, and he shall have abundance; but from him that hath not shall be taken away even that which he hath". Derived from the Gospel according to Mathew 25:29 and presumptive evidence from Nobel laureates, the eminent sociologist Robert Merton suggests that one of its consequences in science communication is that it may serve to heighten the visibility of contributions to science by scientists of greater standing and to reduce the visibility of contributions by authors who are less well known [67]. This may be one contributing factor to producing "sleeping beauties"—academic papers whose importance is not initially recognised for a number of years after publication, and whose citation history by peers exhibits a long hibernation period followed by a sudden spike of popularity [68].

By synthesising evidence and recommendations from primary social science research, *Climate Outreach* has produced a handbook that sets out six principles that can be applied more generally to science communication. Box 2.1 is an annotated summary of these principles.

Box 2.1 Six Principles for Science Communication

1. **Be a confident communicator:** *Scientists are generally highly trusted. By using an authentic voice, you can communicate effectively with any audience.*
2. **Talk about the real world, not abstract ideas***: Although they define the science and policy discourse, "big numbers" don't relate to people's day-to-day experiences. Start your conversation on common ground, using clear language and examples your audience is more likely to be familiar with.*
3. **Connect with what matters to your audience***: Research consistently shows that people's values and political views have a bigger influence on*

their attitudes than their level of scientific knowledge. Connecting with widely shared public values, or points of "local interest" in your communication and engagement makes it more likely that your science will be heard.

4. **Tell a human story**: *Most people understand the world through anecdotes and stories rather than statistics and graphs, so aiming for a narrative structure and showing the human face behind the science when presenting information will help you tell a compelling story.*
5. **Lead with what you know:** *Focus on the "knowns" before the "unknowns" and emphasise where there are areas of strong scientific agreement around a topic.*
6. **Use the most effective visual communication:** *Choosing images and graphs is just as important to do in an evidence-based way as verbal and written communication.*

Adapted from [69].

Despite the six principles in Box 2.1, neither the absence of evidence, nor the absence of medium-term consistency in setting targets is conducive to meeting the required seventeen Sustainable Development Goals. Currently, only about 20% of countries have data for *Goal 13: Climate Action* [70] and the use of multiple "due by" target dates from multiple sources undermines both effective science communication and broader societal acceptance.

A Language Game to Explore Complexity

An essential part of rationality is dealing with randomness in our lives and uncertainty in our knowledge—Steven Pinker

The prickly polymath, Nassim Taleb, in his book *Antifragile* introduces the neologism of its title and the concept of "antifragility", which he summarises as being those phenomena where there is more upside than downside from stressors and shocks. His most persuasive example is the contrast that he draws between non-biological and biological systems. The former are mechanical and non-complex whereby stressors cause material fatigue, while the latter are organic and complex in which the absence of stressors causes atrophy [33, p. 59]. In other words, stressors are beneficial for the maintenance of biological systems, but harmful to mechanical systems. The concept of antifragility is not fundamentally different from adaptation in ecology, but it is novel when applied to non-living systems. Adaptation is briefly addressed in *Antifragile* in a footnote wherein it is stated that "organisms can only have non-anticipative strategies – hence nature can only be nonpredictive"

[33, p. 68]. Conversely, humans have agency and can attempt to create the conditions for adaptability to take hold, although the final outcome cannot be accurately predicted. In addition to his conceptual leap, it seems that Taleb has also engaged in a language game which has borne fruit. We will follow suit by applying the same simple technique of affixing the prefix "anti" to related and existing concepts as shown in Table 2.1. Linguists describe this as a "polysynthetic" approach, which denotes that with each additional affix another layer of story accrues [14]. This exercise is not trivial. For example, let us try to ascertain the impacts of farming methods along a continuum from resilient to *"anti-resilient"* as if we were a Swedish agricultural policymaker. Swedish scientists carefully adding energy inputs versus outputs have been able to show that a tractor needs 67% more energy than the feed required for a draft horse tilling the same field. The tractor is far more effective than the horse; it will produce almost two and a half times the crops from the same area that the animal's ancestors did in the previous century, but it needs 13 times as much energy [71, pp. xiii–xiv]. Let us assume for the purpose of this exercise that there is also an extreme weather event that destroys a season's production for both small-scale and industrial Swedish farmers in a given area.

Both types of farmers can recover more quickly if they deploy their tractors, but their relative capacity to adapt to the financial and other effects of the shock will vary. The small-scale farmer is more likely to be courting disaster because of the high cost of tractor fuel compared to traditional horse feed. He may therefore choose, by reverting to his horse, to *dissipate* his overall risk exposure by "buying time" in the expectation of more favourable weather conditions to come. In this scenario, the hypothetical large-scale farmer has *absorbed* the weather shock and proved to be resilient; his small-scale counterpart has dissipated the same weather shock and proved to be *"semi-resilient"*, that is, was able to respond only to its immediate effects. This simple analysis arose simply from the look-up table above that was created from what are largely antonyms, and more specifically, by contrasting *"absorbs shocks"* to *"dissipates shocks"*. Similarly, we can also imagine *"anti-robust" and "anti-resilient"* situations, all stemming from a look-up table designed to explore the semantic implications of some common terms from complexity

Table 2.1 A language game to explore complexity

Initial condition	Risk condition	Discrete positive/negative impact
Fragile	Susceptible to shocks	Negative
Antifragile	Adapts to shocks	Positive
Robust	Resists shocks	Positive
Anti-robust[a]	Succumbs to shocks	Negative
Resilient	Absorbs shocks	Positive
Semi-resilient[a]	Dissipates shocks	Positive
Anti-resilient[a]	Unable to absorb or dissipate shocks	Negative

[a]New language formulation

science—in other *words*—we played a language game. If we accept that as suitable a word as "antifragile" was not available to Nassim Taleb, we can conclude that he brought the concept out of the mystical and into the epistemological, and certainly in applying adaptation-like qualities to non-living things. Knowledge grows in this way—by applying one principle from one domain to a completely different domain. A less generous interpretation of the value of antifragility is dependent on whether or not the analogy with adaptation is sound. If so, then the main contribution of Taleb's bestselling book would be dialectic in characterising "antifragile" as something that is continuous, whereas "adapted" implies an end-state, something that is discrete. Similarly, "adaptable" implies something that is to come in the future, but "antifragile" implies something ontological, something that already exists. But the essence of these distinctions is only brought about by human language itself given that adaptation occurs in nature irrespective of whether it is emergent, or is directed, or what it is called. If the same can be said about non-living things, then Francis Bacon's earlier refrain rings true about equating words with reality.

Fragility, resilience, and similar concepts influence our understanding of sustainability. Although "sustainability" has become a popular refrain, it is not a well-defined term in the literature and is moreover often used interchangeably with "sustainable development" as introduced by the 1987 World Commission on Environment and Development (Brundtland Commission). According to the Commission, "sustainable development is development that meets the needs of the present without compromising the ability of future generations to meet their own needs". This is currently the most popular definition with its inferred emphasis on human development—not environmental stewardship. Consequently, sustainability needs to be better defined such that it is understood to be "the development that meets *all* of living Earth's needs now and in the future". At base, the current collective action problem regarding global ecological concerns is rooted in this lack of a generally accepted definition of "sustainability" because multiple competing definitions subtract attention and diffuse thinking on solutions. Thought comes from memory and past knowledge, whereas thinking implies the present, some activity which may include critical sensitivity to what can go wrong, perhaps interspersed with new ideas and solutions [43, pp. 60–61]. Put differently, we have to move from past sustainable development thought to present-day *sustainability thinking* that integrates all Earth-bound things into one living Earth—a reinvigorated *Gaia* [72]—the Earth goddess.[1] Such an existential requirement implies that living organisms as dynamic phenomena must actively survive and reproduce in a goal-directed sense [73]. If life itself is the manifest purpose of evolution, then purposive agency needs to be brought back into the study of living organisms [74]. We can choose to initiate forms of thinking and behaving that speed up our evolution by many orders of magnitude and shift them away from the blind forces of necessity and chance in directions of our own choosing according to our own values [5, p. 124]. It is important to grasp the political reorientation that could emerge if we were to objectively conceive of

[1] In Greek mythology Gaia was the personification of Earth, see https://en.wikipedia.org/wiki/Gaia

knowledge as encompassing all the activities necessary for our own existence. The nineteenth century was the era of the social question; the twenty-first of the *new geo-social* question [75].

The existing context is already clear—increasing complexity. In the age of AI, computers, by gaining command of language are seizing the master key unlocking the doors of all institutions. We use language to create not just legal codes and financial instruments but also art, science, and religions [76, p. 208]. Artificial intelligence will in time bring extraordinary benefits to medical science, environmental issues, clean-energy provision, and many other areas. But precisely because artificial intelligence draws conclusions regarding an evolving, as-yet-undetermined future, ambiguity and uncertainty are inherent in its results [77].

Chapter 3
Shifting Knowledge Boundaries

> *Intoxicated by the achievements of his mind, by modern science and technology, he forgets that his knowledge has limits and that beyond these limits lies a great mystery, something higher and infinitely more sophisticated than his own intellect*—Václav Havel

Marcelo Gleiser, in his book *The Island of Knowledge* has coined a metaphor of the same name to describe humanity's accumulated knowledge as an island surrounded by a vast ocean of the unknown which hides innumerable mysteries. We can imagine the Island of Knowledge growing as we discover more about the world and ourselves, the coastline marking a rugged boundary between the known and unknown. We would expect that the more we know of the world, the closer we would be to some sort of final destination, but the island's growth holds a surprising but essential characteristic. We see that as the Island of Knowledge grows, so do the shores of our ignorance, the boundary between the known and the unknown. Learning about the world doesn't bring us closer to the hope for destination of truth, but to more questions and mysteries. The more we know, the more we realise the extent of our ignorance and want to ask and know even more [78, pp. xxi–xxii]. Lessons can also be drawn from Nobel Prize winners who, though sensitive to the evidence provided by the work of others in their respective fields, tend to be self-directed men and women moving confidently into new fields of inquiry once they are satisfied that a previous one has been substantially exhausted. In these endeavours they demonstrate a high degree of venturesome fortitude and are prepared to take on important though difficult challenges rather than settle for easy and secure ones [79]. Psychologist Mihaly Csikszentmihalyi in his study of eminent scientists like Linus Pauling and Jonas Salk concluded that what differentiated them from their peers was their cognitive flexibility, and as the occasion required, their willingness to shift from one extreme to the other. Similarly, the same pattern held for great artists and also supported by a study of highly creative architects [19, p. 27] (Fig. 3.1).

Some forays into the unknown can be made by embracing randomness without an overreliance on the scientific method. Coupled with good luck, explorations of this kind may lead to highly consequential outcomes. The field of medicine offers many such instances from the discovery of penicillin to the fortuitous multiple repurposing of aspirin. Outwards from what is known, the best way forward according to Nassim Taleb is *nonteleological tinkering* because human innovations drift

E. Bellos, *Sustainability Thinking*, https://doi.org/10.1007/978-3-032-11244-6_3

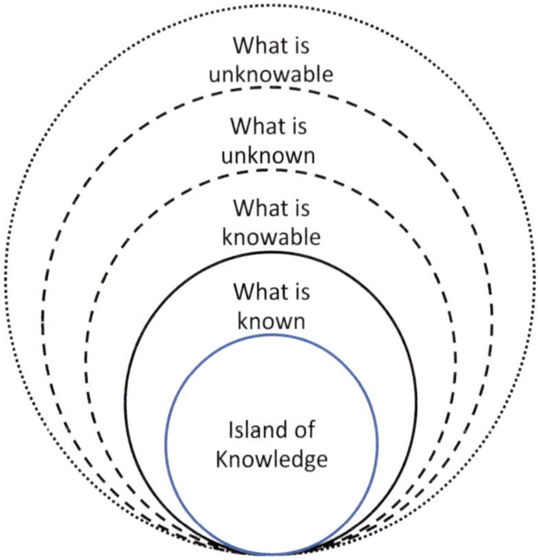

Fig. 3.1 The shifting boundaries of the Island of Knowledge. (Source: Author)

and one needs savant-like abilities to keep capturing the opportunities that arise as opposed to being moribund by bureaucratic thinking. Payoff from research should correspond to the number of trials undertaken, not the total funds expended on them. The right approach requires a certain faith in blind funding, what is called the "one divided by N" policy, meaning small expenditures per trial but lots of trials—broader than would normally be expected. Doing so preserves the option to capitalise on a potentially huge payoff.[1] And who is best placed to fund lots of research and lots of trials? Governments, companies, and citizen groups are—singly or jointly. A kind of "medical DARPA" (the Defense Advanced Research Projects Agency of the United States) which seeded the Internet and GPS could fund health research that leads directly to medical products. In this scenario, open-source initiatives could also lead to innovation as long as they enjoyed sustained financial support. The response to COVID-19 has been more open-source than any drug development effort to date. COVID-19 has invigorated the call for a "Linux for drug research and development", invoking the success of the famed open-source operating system. Under this model, researchers, scientists, and doctors—who sometimes find novel uses for drugs by prescribing them to address conditions other than their original purpose—could readily contribute to various research projects. A 2018 study from the Cambridge Judge Business School on repurposing medicine found that about 31% of drugs approved by the FDA had the potential to treat a similar illness to the original target disease with a 67% success rate [80]. Such repurposing would help to counter the multiplying number of novel drug interactions with each other, which would reduce overall risk from deleterious drug interactions [33, pp. 229–233].

The more we venture beyond the first perimeter of the Island of Knowledge, the more we realise the extent of our ignorance, wanting to know even more and to ask

---

[1] The obverse is known in risk management as low probability, high impact events.

even more questions. The island's outer limits are the shore of philosophy, which asks questions such as "what is consciousness"? Philosophers by repeatedly asking, but without the means of answering, nevertheless set the stage for *protoscience*, an incubation period that allows science to catch up and to answer back. The history of philosophy tells us that several difficult questions have eventually been answered by science [57, pp. 28–29] taking us from the shore of the unknown into that of the knowable. From there, towards safe harbour, the map is well laid out by The Edge Foundation's motto which reads: "to arrive at the edge of the world's knowledge, seek out the most complex and sophisticated minds, put them in a room together, and have them ask each other the questions they are asking themselves". The foundation has been operationalising its motto since 1999 by asking the Edge Annual Question. Notable past questions include: what do you believe is true even though you cannot prove it? What is your favourite deep, elegant, or beautiful explanation? And what scientific idea is ready for retirement? Similarly, the prestigious journals *Scientific American* and *Nature* jointly published a special report in 2018 on the "Biggest Questions in Science". Their selection included: What is consciousness? How did life begin? and How much can we know? Similarly, in October 2018, the UN Committee on Economic, Social and Cultural Rights released a list of 29 questions in its ongoing efforts to define the human right to science. Perhaps most interestingly: what is scientific knowledge and how should it be differentiated, if at all, from traditional knowledge? [61]. The list of questions is potentially endless and proof positive of humanity's limitations with respect to knowledge and wisdom. Here are a few more indicative examples:

- What is reality?
- What is truth?
- Are we alone in the universe?
- When will the Earth System reach its final tipping point?
- Why are we so good at accumulating information and power, but not in acquiring wisdom?[2]
- Why do we have the particular moral beliefs that we do?[3]
- What is the best economic system for the twenty-first century?
- What is the best political system for the twenty-first century?
- Will artificial intelligence mostly benefit or harm society in the twenty-first century?

Not surprisingly, the bigger a fundamental question is, the less we can expect in terms of a definitive answer. Add the psychological tendency to substitute an easier question for a harder one, and getting satisfactory answers becomes even more difficult [82]. But asking people questions may motivate them to rethink their conclusions. Different kinds of counterfactual questions invite people to explore the origins of their own beliefs and to reconsider their various stances. If you get people to

[2] Drawn from [76, p. xii].

[3] Drawn from [81].

pause and reflect, they might decide, for example, that group stereotypes do not apply to individuals. Research suggests that they are more similarities between groups than we acknowledge, and typically there is more variety within groups than between them [19, pp. 136,139]. To counter these tendencies, estimating the outer limits of a knowledge domain can strengthen coarse-grained scientific validity. Without such estimates, relevant scoping, accompanying transparency, and claims of objective truth will be unnecessarily uncertain.

Cognitive Biases and the Biology of Mind

The first rule of the Dunning-Kruger club is you don't know you're a member of the Dunning-Kruger club—David Dunning

There are over 100 identified human cognitive biases of which, arguably, the Dunning-Kruger effect is the most pernicious. This is the tendency of unskilled persons to overestimate their own ability, while its counterpart in skilled persons is the tendency to underestimate their own expertise [83]. It is a contributing factor to what has been dubbed the "post-truth" world and is most pronounced in the lowest performers who tend to mistakenly inflate their own abilities the most [64]. One of the most humbling findings of twentieth century psychology is that a simple regression equation usually outperforms a human expert. Moreover, when an expert who is given the equation and allowed to use it to supplement his or her own judgement often does worse than the equation alone. Even the most powerful predictor beyond the best regression equation—the genome of identical twins—who can be studied in the same home environment and culture provides a typical correlation between the two twins of 60%. Although higher than chance, but way lower than 100%, this means that a lot of variation rcmains mysteriously unexplained. Despite near-identical predispositions, behavioural outcomes are far from being identical: one twin may schizophrenic, the other typical; one twin may be gay, the other straight. Similarly, a comprehensive study of families found that the idiosyncratic traits of each family overwhelmed generic predictors such as income, no matter how carefully the predictors were combined for purposes of the study. Resorting to the simplistic explanation that human behaviour is a combination of nature and nurture is thus not very helpful [84, pp. 279–281].

In his book *Thinking, Fast and Slow*, the late Daniel Kahneman suggests that decision-makers fall victim too easily to the *planning fallacy*. This is because plans and forecasts are unrealistically close to best-case scenarios, and statistics of similar cases are not adequately consulted. The planning fallacy is a mid-range manifestation of optimism bias, which may be thought of as being its upper boundary, while *excessive subjective confidence* partly bounds it at its lower end. Confidence is a condition of decision-makers which reflects the coherence of their information and the cognitive ease of processing it. Declarations of high confidence mainly tell us that the decision-maker has constructed a coherent story in his or her mind, not necessarily that the story is true. Consider the weapons of mass destruction fiasco that led to the Iraq War of 2003 during which no such weapons were found. Without positive countervailing forces leaders may make decisions based on unrealistic optimism instead of the rational weighting of probabilities, gains, and losses. They will underestimate costs and overestimate benefits; they will manufacture scenarios of success while overlooking the potential for miscalculations and mistakes [85, pp. 249–253]. Conversely, successful forecasters often answer questions in a roughly similar way by: (1) unpacking the question into components; (2) marking clear boundaries between the known and unknown in addition to scrutinising assumptions; (3) adopting the outside view and putting the problem into a comparative perspective; (4) adopting the inside view with a focus on the uniqueness of the problem; (5) by exploring similarities and differences between views, including those of prediction markets and synthesising all these different views into a single vision; and (6) by expressing conclusions using a finely grained scale of probability [22, p. 153].

Most of us are terrible at weighing risks presented as abstract probabilities and we discount heavily the well-being of future generations and our future selves, as well as the well-being of distant strangers. We are highly susceptible to conspiracy thinking and display an impressive capacity to deceive ourselves before doing the hard work of deceiving others, and do so in ways that are both psychologically strange and, in a modern environment, ethically indefensible. These human predispositions likely endowed our ancestors with certain advantages, but they also suggest that our species is not biologically prepared for seeking a precise understanding of the world as it actually is [29]. We are still at the beginning of a growing understanding of how animals and humans are able to fashion and use evaluative-causal models to respond efficiently and effectively in the face of emerging challenges. Due to research in psychology and neuroscience, underpinned by a solid foundation in systems theory and philosophy, for the first time we are getting a unified account of how this might actually work. Human intuitive (tacit) thinking has been proposed to be nonverbal, irrational, associative, and emotional in contrast to deliberative (explicit) thinking which is said to be verbal, logical, rule-based, and neutral. This bifurcated picture of two minds, if not completely wrong, is misguided because the difference between intuition and rational deliberation is much more subtle. It amounts to the direction of the "mind's eye". During our ordinary interactions with the environment, intuitive affective (feeling) systems scan the world to supply ongoing automatic guidance of action. During deliberation, however, intuitive affective systems are called upon once again but this time the "gaze" of these systems is directed away from the actual situation and instead towards episodic prospective representations constructed in the mind [2, p. 71], [86, pp. 89, 98].

Are we then helpless victims to "fast" and simplistic thinking? Increasingly less so. For example, engaging with the video platform "Macbeth" (Mitigating Analyst Cognitive Bias by Eliminating Task Heuristics) participants can be trained to recognise cognitive biases and to measure their mitigation over time. Modelled on the board game *Clue*, Macbeth lines up a series of suspects together with the information that will help players determine who committed the crime. Players must then decide whether that information was affected by cognitive biases [87, p. 237].

Invisible Noise Plus Biases

Error in a single measurement = Bias + Noisy Error—Kahneman, Sibony, and Sunstein

Daniel Kahneman (together with Olivier Sibony and Cass Sunstein), in their book *Noise: A Flaw in Human Judgement*, asserts that cognitive biases are well-studied but constitute only half of the problem of errors in judgement—the other half being *noise*. The invisibility of noise is a direct consequence of the human tendency for causal thinking. We naturally pay attention to the particular, following and weaving causally coherent stories about individual cases, in which failures are frequently attributed to errors, and errors to biases. The ease with which bad judgements can be explained away leaves no space for the role of noise in our accounts of errors, rendering noise invisible. *System noise* can be decomposed into *level noise* and *pattern noise* when the same individuals evaluate multiple cases. The former is the variability from the average level of judgements of peer decision-makers, while the latter is the variability of a decision-maker's judgements compared to his or her peers due to the type of cases being considered. At the group level in governments, business, and elsewhere, information cascades and group polarisation can lead to wide disparities among deliberating groups trying to solve the same problem. Information cascades result from uncritical acceptance of early and prevailing viewpoints while group polarisation accentuates one's original inclinations, whether right or wrong, due to a variety of social influences. The potential dependence of outcomes on a few individuals—those who speak first or those who have the largest influence—is particularly problematic. Level noise and pattern noise produce differences among the opinions of group members that are larger than expected or wanted. Moreover, occasion noise (an element of pattern noise), such as fatigue and mood, may affect the judgement of the first person who speaks. Group dynamics can then amplify this and other sources of noise. Consequently, deliberating groups tend to be noisier than statistical groups which merely average individual judgements—provided those individual judgements are derived independently without any "herding" influences [82, pp. 99–106, 219].

Kahneman's and his co-authors suggest that decision-makers, normatively speaking, should strive for a judgement *process* that would produce the best results over an ensemble of similar cases. By focusing on the process of judgement they hope to be able to measure both verifiable judgements and non-verifiable judgements. A verifiable judgement is an *outcome* that can be contrasted against an initial prediction, while a non-verifiable judgement would have neither a benchmark nor a clear outcome [82, pp. 49–51]. This brings us back to the issue of evaluating the impact of particular decisions, more likely than not to be unique and momentous, but without the benefit of comparison with an "ensemble of similar cases". In other words, the outcome and the process of decision-making cannot be easily separated from each other and precaution is called for when doing so. Just the process of decision-making itself is complex and dynamic enough, for example, when new information or evidence is brought to bear on a decision without the challenge of also evaluating the decision's contemporaneous impact. In academic papers, the so-called sleeping beauty phenomenon can have an outsized, delayed, and silent influence on group deliberations. It can be hypothesised therefore that embedded information, whether on paper or otherwise, may generate critical emergent conditions rather than mere background noise and should accordingly inform the further study of deliberating groups. Undoubtedly the study of errors associated with psychological heuristics is important given their potential negative consequences, such as when mental shortcuts are overly relied on; however, *adaptive heuristics* often lead to enhanced outcomes. Heuristics generate the adaptive prospecting that reveals possible futures. At the margins, heuristics make errors, but for the most part, they are our first and most robust way of exploring the future. When they do not lead to biases, positive adaptive heuristics allow for goodness, beauty, rightness, and truth to emerge over and above their negative counterparts [88, p. 320].

The Practice and Malpractice of Science

> *I couldn't claim that I was smarter than sixty-five other guys—but the average of sixty-five other guys, certainly*—Richard Feynman

Being a scientist is not just a profession; it is a mode of thinking when searching for the truth. Thinking like scientists involves more than just reacting with an open mind—it means committing to active effort to be open-minded. It requires searching for reasons why we might be wrong, not for reasons why we must be right—and revising our views based on what we learned. Science is reiterative. We can run trials (experiments) to test ideas (hypotheses) and ask questions to increase our knowledge. Hypotheses have as much of a place in our lives as they do in the laboratory; nor are scientific tools reserved for people with fancy titles in ivory towers. As discerning consumers of scientific information, we can influence practicing scientists from becoming preachers when they present their conclusions as gospel and dismiss constructive criticism; or when they veer into politician terrain swayed by popularity rather than accuracy; or when they become like prosecutors unaccepting of challenges instead of concentrating on discovery [19, pp. 20, 22, 25, 86].

To varying degrees and success science moderates competing interests and thereby creates the solution space for progress to be made. Consider that it may be the case that fewer people have been directly influenced by Darwin's theory of evolution than by its indirect ideological ramifications in the form of social Darwinism [89, p. 22]. Figure 3.2 depicts the conditions and ideal role of science in creating consensus. Science is good at identifying problems that can be solved within a prevailing knowledge ecology and making a case for their resolution. Climate change is the current preeminent example yet polls indicate that there are at least six camps of thought among the American public ranging from dismissive (10%) to alarmed (31%). Presenting more nuanced points of view by scientists would actually increase credibility and hence acceptance. As consumers of information, we can also play a

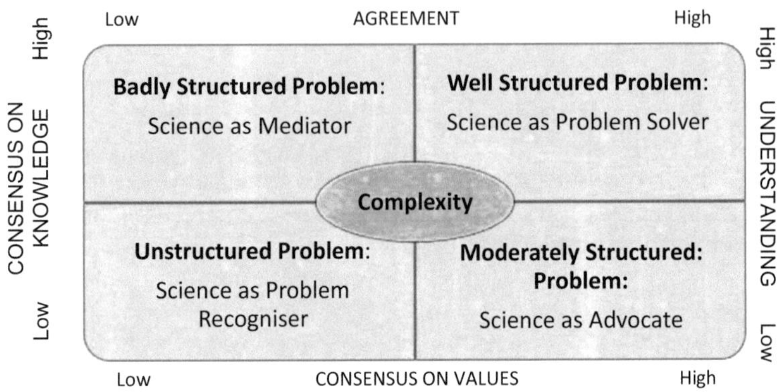

Fig. 3.2 The solution space of the science commons. (Author based on [90, p. 395])

role by favouring content and sources that include many sides of an issue rather than a binary oversimplification [19, pp. 170–171].

Science is not as good at consensus building when values such as equity are at issue, for example, what role intellectual property should play in worldwide vaccine distribution. Recently, the promise of "big data" has cast the sciences into a sea of amorphous information with few constraints on where it might drift. The difficulty with this way of doing science is that for any large body of data relating to a complex problem with many variables, the number and strength of possible causal links among the relevant variables is inestimably larger than the number a scientist could actually think up and test. One cumulative result of these converging stresses is the well-recognised pervasive bias towards the new result. If biases were random, then they would be able to cancel each other out over multiple studies which is not the case due to powerful forces that encourage positive results—the transformational, attention-grabbing findings—something that announces you as part of the elite [91]. The individual scientist has three common pathways open to him or her: first, and most common, is a diminishing emphasis on research; second, an overconfident diversification into other fields; and third, to continue to do what one is competent at and accept that some new techniques are more suitable to the young. The third option of aspiring to be on a plateau rather than scaling new heights may be the preferred and most society-friendly option [92].

The issue of research quality was illuminated in May 2016 by a survey of more than 1500 scientists published by the prestigious journal *Nature* which showed that 80% or more of respondents believed that the scientific enterprise was being undermined by such factors as inadequate peer review, selective reporting of data, poor statistical analysis, pressures to publish, and insufficient attention to replication [93, p. 20]. John Ioannidis foresaw the replication crisis in 2005 and shook up his medical colleagues by his article *Why Most Published Research Findings Are False*. At issue in this instance is that many phenomena that biomedical researchers focus on are interesting but are a priori unlikely to be true, requiring highly sensitive methods to avoid false positives, while many true findings, including null results and successful replication trials, are considered too boring to publish [84, p. 162]. And from personal experience I can recount that, although different academic journals employ different quality assurance procedures, those that use peer review—a key step on which legitimacy relies—is less than satisfactory. The self-selection of unvetted reviewers (or if vetted, with vetting status not disclosed) may introduce unknown biases into the process. It was recently reported that some peer reviews plagiarise other reviews which may be a part of a widespread problem stemming from the lack of confidence of secondary reviewers, as well as time pressures [94]. Moreover, submitted materials from authors that remain unpublished may not be sufficiently credited if reviewers choose to use those materials to then successfully publish their own academic work. In part this is because plagiarism software used to check authenticity can only do so for articles held out *and* actually published. Perhaps the moniker "peer review" says it all—it is review by other scientists—review that is self-referencing, circular, and not open to outsiders. Adam Grant recounts a funny episode in which a reviewer in rejecting his manuscript suggested that he go back and read the work of Adam Grant! [19, p. 86].

Clearly, several reforms would be in order. Academic journals might want to consider establishing an equivalent model to that of journal editorial staff whereby the qualifications, formal affiliations, and areas of expertise of both editors *and* reviewers are published with the additional provision that the specific identity of individual reviewers of a submitted paper remains undisclosed from among the total set of *vetted reviewers*—including non-scientists—for those journals that opt for single- or double-blind reviews. Journal rankings should then positively account for those publications that adopt the proposed scheme. A non-traditional model of quality assurance was introduced recently by the publisher *Public Library of Science* which manages some of the highest impact journals in biology and medicine, including *PLOS ONE*, a journal that covers all the natural sciences. Although still based on peer review, pre-publication referees are asked to evaluate a submitted academic paper purely on the basis of whether the work was carried out in a sound scientific manner. The paper is then published and judgements about its significance, relevance, and impact are made by means of online community responses. The model has proved to be very successful and has recently been adopted by the Nature Publishing Group with its launch of *Nature Scientific Reports* [95]. Similarly, the state of the art in pre-publication scientific methodology is to "pre-register" the details of a study in a public registry that reviewers and editors can check for any *post hoc* manipulation [84, p. 146].

What is more, we have become overly dependent on governments. Modern-day governments have in turn become overly dependent on externally produced knowledge although there are real limits to the relevance and usability of knowledge across organisational and other contexts. When there is an overabundance of information, it may be more important to know which knowledge to include in decision-making, rather than having information for its own sake. Collecting evidence—when full diagnosis of a problem may be too resource intensive or even impossible—is a significant challenge. Sufficiency of information could be permanently out of reach and should be acknowledged as such. When working on problems related to broader systems or highly complex problems, there is often no definitive answer [96, p. 21]. "Post-normal science", so named as to indicate a departure from the curiosity-driven or puzzle-solving exercises of normal science, accepts uncertainty and storytelling as crucial elements in the process of decision-making. Normal science has been successfully extended from the laboratory to the conquest of nature through applied science. However, the traditional scientific approach is no longer appropriate for the solution of sustainability problems. The social, technical, and ecological dimensions of sustainability problems are so deeply interconnected that it is impossible to consider them separately, one at a time, as conducted in the reviews and analyses of conventional disciplinary fields [97, p. 129]. Research must become practically implementable to support decision-making through "achievable visions" that are co-developed with users to a sufficient level of complexity. There are many existing areas of global change research that can be developed further whenever appropriate by integrating natural and social science. Attaining a sustainable world will require solution-focused research that bridges disciplines and engages stakeholders, helping to build the consensus needed for effective action at national, regional, and global scales [98]. Broad coalitions of citizens, businesses, non-profits, and government agencies have the potential to transform how we act as stewards of the biosphere. Science has an important renewed role to play in this context as an honest broker, informing evidence-based action, and in the coproduction of knowledge [99].

The international scientific community has a responsibility to society to exhibit more leadership. This should involve a much greater focus on *strategic applied research*, namely, good science that is also squarely directed at understanding the contextual challenges that decision-makers face so as to simultaneously provide solutions to these. This will require global science to embrace wider and stronger partnerships with decision-makers, while maintaining a foundation in basic research to explore the "known unknowns". In such a situation, the transparency achieved would be equally as important as the quantity, quality, and relevance of the underlying facts, methodologies, and framing strategies under consideration. The wise consumer of scientific reports and science journalism would be wise to "read the label"—both good and bad ingredients are likely to be found. As citizen philosophers, we should be requesting, if not demanding, greater initial and ongoing transparency.

Chapter 4

Sustainability and Other Global Existential Risks

Can the global village cope with its village idiots—especially when even one could be too many—Martin Rees

Short of a cosmic catastrophe, not hastening on a course towards environmental sustainability poses the greatest medium-term risk to humanity. Guided by *precautionary optimism*, a healthy tension between stability on the one hand and dynamism on the other may bring us to a beneficial state at "the edge of chaos". As Britain's Royal Society has pointed out, "objective" risk is measurable from data, whereas "perceived" risk is a subjective human response to danger. In the broadest terms, perceived risk involves categorising risks as acceptable, tolerable, or intolerable (traffic light model), which in turn depends on the relative weight of values, as well as the nature, extent, and probability of the risk consequences as assessed and evaluated by those entitled to make these decisions [100, p. 178]. Consequently, risk literacy is essential for policymakers, doctors, judges, and others who hold our lives in their hands [84, p. 171]. The speed and magnitude of global change, the increasing intra and interconnectedness of the social and natural systems at the planetary level, and the growing complexity of societies and their impacts on the biosphere result in a high level of uncertainty. Uncertainty and complexity are two of the three elements of risk, the third being ambiguity [100].

Nick Bostrom's definition of existential risk is one where an adverse outcome would either annihilate Earth-originating intelligent life or permanently and drastically curtail its potential. In his opinion setting the probability of existential threats lower than 25% would be misguided, but even if the probability were much smaller (say, ~1%) the subject matter would still merit extremely serious attention because of how much is at stake. The best current subjective estimate of what the objective risk factors might be should serve as our guide. This requires foresight and a willingness to take decisive preventive action and to bear the moral and economic costs of such actions [101]. Although there may be a very high value in studying existential risks and in analysing potential mitigation strategies, it is striking how little academic attention these issues have received compared to other topics that are less

E. Bellos, *Sustainability Thinking*, https://doi.org/10.1007/978-3-032-11244-6_4

important. There are unresolved foundational issues, particularly concerning observation selection theory and population ethics, which are crucial to the assessment of existential risk, all of which are compounded by psychological factors that make it difficult to think clearly about issues such as the end of humanity. Many factors conspire against the study and mitigation of global existential risks. Research is inhibited by the multidisciplinary nature of the problem, as well as the need for new scientific research methodologies. Multidisciplinary and epistemological challenges, academic distractions and diversions, cognitive biases, free rider problems, moral lethargy, scope-insensitivity, institutional incompetence, and the political exploitation of unquantifiable threats are some of the barriers to effective resolution. If more resources were available to research existential risks, there is a strong possibility that they would mainly flow to the relatively minor risks that are easier for some established disciplinary community to study using familiar methods, at the expense of far more important risk areas—machine superintelligence, advanced molecular nanotechnology, threats related to the simulation-hypothesis, future advances in synthetic biology, or totalitarianism—which would require a radical shift in research focus. Or it could be that research is directed towards lesser global catastrophic risks, such as nuclear war, which probably does not pose an outright existential risk [102].

In his paper *Existential Risk Prevention as Global Priority*, Bostrom points out that humankind is introducing entirely new kinds of existential risk—threats we have no track record of surviving for which our longevity as a species offers no prior grounds for optimism. Consideration of specific existential risk scenarios supports the notion that the great bulk of threats in the foreseeable future will consist of anthropogenic existential risks especially in light of the international nature of many of the required countermeasures and the shared stakeholdership of all future generations. In particular, most of the major risks seem to be linked to potential future technological breakthroughs that may radically expand our ability to manipulate the external world or our own biology. It follows that as our powers expand, so will the scale of their potential consequences—intended and unintended, positive and negative. There will be some types of putative existential risks for which the main uncertainty is normative and others where the main uncertainty is descriptive, but in either case, if something is not known to be objectively safe, it should be assumed to be risky, at least in the subjective sense relevant to decision-making. In consequence we would want to reach a state in which we have (1) far greater intelligence, knowledge, and sounder judgement than we currently do; (2) far greater ability to solve global-coordination (public good) problems; (3) far greater technological capabilities and physical resources; and (4) values and preferences that are not corrupted in the process of getting there. Factors 2 and 3 expand the option and solution sets available to humanity; factor 1 increases our ability to predict the outcomes of the available options and understand what each outcome would entail in terms of the realisation of human values, and factor 4 makes humanity more likely to want to realise human values in the end. But speculative risk mongering could be exploited to rationalise self-serving aggressive action, expansion of costly and potentially oppressive security bureaucracies, or restrictions of the very same civil liberties that keep societies free and sane. The net result of false approximations to the rational ideal could easily be a net increase in existential risk. A reactive approach to existential risk to observe what happens, limit damages, and then implement improved mechanisms to reduce the probability of a repeat occurrence does not work when there is no opportunity to learn from failure [102].

The World Economic Forum's 2024–2025 *Global Risks Report*, its 20th edition,[1] notes that over two decades, *environmental risks* have steadily consolidated their position as the greatest source of long-term concern. Also, a sense of alarm has been reported as increasing in the short term. From pollution to extreme weather, environmental problems were deemed to be upon us and the need to implement solutions urgent. Two-years out extreme weather events were ranked second in potential risk severity after *misinformation and disinformation*. Ten years out, four of the five top risks were considered to be environmental, respectively: extreme weather events; biodiversity loss and ecosystem collapse; critical change to earth systems; and natural resource shortages. The fifth risk over this period was misinformation and disinformation. Across time, from the two to the ten year periods, *inequality* persisted unchanged as a risk and was perceived to be the most central by contributing to diminished trust and the weakening of our shared values. Interestingly, *economic risks* did not make it to the top ten ranking either in the two or ten year time horizons. *Is a social phase transition of our collective attention in the offing from pure economic to more pressing environmental concerns?* The edited volume *Global Catastrophic Risks* covers everything from global warming to comet impacts but has a blind spot regarding non-sustainability as the driver of many subsidiary risks. Conversely, a comprehensive treatise such as *Sustainability or Collapse* covers sustainability but fails to view it as an overarching global risk driver. Some progress has been made in identifying the tension between energy security and energy transition as a country-level systemic risk [103], but this has not yet been extended to a global scale and broadened to include sustainability. There is a huge gap in the literatures and therefore in our knowledge and needed attention concerning *sustainability itself as a global existential risk.* At least climate change has been acknowledged as posing a quintessentially universal risk to all nation-states. This risk is of an urgent character requiring the selection of a corresponding standard of conduct subject to the principle of common but differentiated responsibilities and respective capabilities [104, p. 50].

Having defined existential risk as an adverse outcome, Nick Bostrom has elaborated on his initial definition by recommending that we need not reach some state that is directly "sustainable" in the sense that we could remain in it for some time. Rather, we should focus on creating a developmental process that offers a high probability of avoiding existential annihilation. The notion of existential risk

[1] The report highlights the latest findings from its annual Global Risks Perception Survey involving over 900 global leaders across academia, business, government, international organizations, and civil society. One hundred thematic experts from the Global Risks Report Advisory Board, the Global Future Council on Complex Risks, and the Chief Risk Officers Community contributed insights.

suggests a new way of thinking about the ideal of sustainability which should be reconceptualised in dynamic terms, as aiming for a sustainable trajectory rather than a sustainable state. An urgent need exists to complement the existing knowledge base of risk analysis and management by developing frameworks and models that would enable multilevel systems resilience analysis, engineering, and management [105]. And more research into existential risks is needed—detailed studies of particular aspects of specific risks, as well as more general investigations of associated ethical, methodological, security, and policy issues. We cannot necessarily rely on the institutions, moral norms, social attitudes, or national security policies that developed from our experience with managing other sorts of risks. Public awareness can also be built up so that constructive political debate about possible countermeasures becomes actionable. Civic awareness of the global impacts of human activities on systems, processes, and risks is increasingly studied from a global perspective by many scholars—environmental scientists, epidemiologists, demographers, economists, and others. In particular, problems such as climate change, cross-border terrorism, and international financial crises direct attention to our global interdependency and threats to the Earth system. In general, the subject of risk seems to have risen in prominence. Given recent advances in knowledge, methods, and attitudes, the conditions for ensuring the epistemic attention they deserve are unprecedentedly propitious [102].

It is also undeniable that human civilisation would not be what it is today without the breakout opportunities that technology has enabled in the course of human development. Witness electricity, computers and computing power, the Internet, photography, antibiotics, vaccines, genetic engineering, materials engineering, mass transit, the green revolution, and the "seeing" technologies we encountered earlier. What we have the power to affect is the *rate* of development of various technologies and potentially the *sequence* in which feasible technologies are developed and implemented in what we might call *differential technological development*, that is, the slowing of dangerous technologies and the acceleration of beneficial technologies—especially those that ameliorate the hazards posed by some other technologies as well as the potential benefit of reducing the threat from natural hazards [101]. Our focus should be on maximising the chances that we will someday attain technological maturity in a way that is not irremediably and dismally flawed. Conditional on that attainment, we may have a reasonable chance of realising our enormous axiological potential mindful that the problem of how to avoid existential risk has no known solution under the present circumstances [102]. And although resilience is not a substitute for sound system design or risk management, but rather a complementary approach that uses strategies of adaption and mitigation to improve traditional risk management, [105] it buys time and its study holds promise for some solutions. For example, a resilience matrix has been proposed by mapping the four basic constructs set out by the U.S. National Academy of Sciences, namely, *plan/prepare*; *absorb*; *recover*; and *adapt* against four system domains comprised of *physical*, *information*, *cognitive*, and *social* factors to produce sixteen elements that can be used as guidelines to formulate resilience metrics [106]. A comprehensive, modern, and, yes, ongoing sustainability research agenda is required to assess the ongoing risks that ignoring or minimising sustainability would bring to a complacent world.

COVID-19 and the Doomsday Clock

Only entropy comes easy—Anton Chekhov

In March 2020 the world was in a crisis due to the COVID-19 virus, a new highly infectious coronavirus that causes respiratory distress mostly in humans up to and including death. Amidst the catastrophic threat to humanity that the pandemic induced it became increasingly clear that global collective action underpinned by a collective perspective was needed. The launch of the COVID-19 Global Science Portal by the International Science Council[2] was an initiative in this direction designed to provide a curated knowledge platform so as to jointly respond to the crisis. About a year into the pandemic, a handful of COVID-19 vaccines had already been developed, albeit with concerns about their equitable distribution across the globe. The effects of the pandemic were examined holistically with some startling insights, most notably that the virus was manipulating human behaviour to enhance its own evolutionary fitness. This resulted in a kind of arms race with the virus attempting to suppress feelings of sickness during times of peak transmissibility so as to enable hosts to go about their usual activities and thereby come into contact with others whom they would infect and propagate the virus. Mood disorders, such as mania, could also temporarily increase activity levels and decrease feelings of sickness during times of peak transmissibility initiating a "tug-of-war" between greater host sociability on the one hand and host unsociability on the other hand; which in the case of the latter, host healing would have been prioritised instead [29]. The COVID-19 crisis put a spotlight on the need for rapid collective foresight to steer the response to this and future global systemic crises with more agile collective foresight that can rapidly draw on the expertise and imagination of a broad diversity of people, knowledge systems, and worldviews to help society anticipate and address risks and opportunities [107]. In the fullness of time many pandemic and risk-related lessons ranging from individual psychological resilience to public health preparedness will arise and are sorely needed as in the case of "Long Covid" and its debilitating effects. In 2025, the first global pandemic treaty under the auspices of the World Health Organization (WHO) was agreed to and awaited ratification by its World Health Assembly. Earlier that year, the United States had withdrawn from WHO undermining its operational effectiveness, but not the signal achievement of obtaining a multilateral agreement among WHO's 194 member states. The final treaty will come into force one month after sixty countries have ratified it [108].

While humanity was engrossed in the pandemic of 2020, the doomsday clock was ticking. Beginning in 1947 and maintained since then by the members of the

[2] https://council.science/covid19/

Bulletin of the Atomic Scientists, the metaphorical Doomsday Clock tracks in "minutes" the likelihood of a human-induced global catastrophe. "Midnight" is when the catastrophe may materialise mainly due to nuclear risk and climate change. The Science and Security Board of the Bulletin also monitors new developments in technology and the life sciences that could cause irrevocable harm to humanity. The Clock's setting in 1947 was seven minutes to midnight. Discussing the change to $2^1/_2$ minutes in 2017, one of the Bulletin's scientists warned that our political leaders must make decisions based on facts and called for action by *wise* public officials and *wise* citizens to try and steer human life away from catastrophe while we still can (my emphasis) [109]. Before the thermonuclear atomic bomb, humanity had to live with the idea of his or her death as an individual—but from then on—humanity has had to live with the idea of its death as a species [10, p. 322]. The clock was moved forward to two minutes to midnight in January 2018 and left unchanged in 2019 due to the twin threats of nuclear weapons and the increasing effects of global warming. On 23 January 2020, it was reset forward to 1 minute 40 seconds before midnight based on the increased threats to human security posed by nuclear blunder, exacerbated by the rate of climate change. In January 2023, the Science and Security Board moved the hands of the Doomsday Clock forward again to 90 seconds to midnight in large part due to Russia's invasion of Ukraine on 24 February 2022. Russia's war on Ukraine raised profound questions about how nation-states interact, eroding the norms of international conduct that underpin successful responses to a variety of global risks. Among other unintended consequences, it jeopardises renewal of the new START nuclear treaty set to expire in February 2026, the last remaining nuclear weapons treaty between Russia and the United States [110].

In 2025, the big hand of the doomsday clock was moved to 89 seconds to midnight—the closest to global catastrophe it has ever been. Ongoing conflicts in the Middle East threatened to spiral out of control into a wider war with no warning. The continuing non-nuclear war in Ukraine may become nuclear unexpectedly due to a rash decision, accident, or miscalculation. Countries with nuclear weapons increased their size, while others are considering developing arsenals of their own thereby increasing the ways in which a nuclear war could start. Systems that deploy artificial intelligence in military targeting have been used in Ukraine and the Middle East. The extent to which machines will be allowed to make military decisions in the future, including the use of nuclear weapons, remains an open question. The rapid growth of artificial intelligence also raised the risk that terrorists or weaker countries may use it to design biological weapons. Furthermore, the long-term outlook for the world's attempts to deal with climate change remains poor as most governments have failed to enact the policy initiatives and financing necessary to curtail global warming. All these dangers are greatly exacerbated by a powerful threat multiplier, namely, misinformation, disinformation, and conspiracy theories that degrade the integrity of the communication ecosystem [111]. Even if granted the requisite authorities, it is hard to envision any AI system being able to timeously and effectively self-correct all types of critical miscommunication, including deliberate miscommunication. Consider the missing human link in the form of Lieutenant Colonel Stanislav Petrov of the Soviet Union Air Defence Forces. Petrov was a crucial human link in an otherwise automated decision chain who in 1983 used his judgement to report a technical malfunction rather than a U.S. missile launch thereby averting a likely nuclear conflagration by overriding his computers' and satellites' erroneous early warnings, which were simply picking up the sun's reflection on the clouds and not a first-strike flare [112, pp. 85–89]. The totality of risks humanity faces may be encapsulated by the term "ecocide"—the antithesis of sustainability given that technological uncertainties and geopolitical emergencies draw attention away from longer-term, but even more life threatening sustainability crises.

Our Biosphere Home

> *Human beings are good at many things, but thinking about our species as a whole is not one of our strong points*—David Attenborough

The thin 20 kilometre (12 mile) life-giving layer of the Earth we call the biosphere is shaped by global atmospheric circulation jet streams, atmospheric rivers, soil formation, movement of tectonic plates, earthquakes, volcanic eruptions, water vapour and precipitation patterns, the distribution of the ozone layer, the spread and retreat of ice sheets and glaciers, the upwelling of coastline currents, and the ocean's global conveyer belt. Across the ocean and the continents, the living biosphere is dynamically interconnected with the broader Earth system and its climate system, atmosphere, lithosphere, cryosphere, and hydrosphere, as well as the carbon, nitrogen, and other biogeochemical cycles that are essential for all life on Earth. It is this complex adaptive interplay between living organisms, the climate, and the broader Earth system that has evolved to be our biosphere [99].

Although the *Netflix* documentary *Breaking Boundaries*, narrated by Sir David Attenborough, has helped to convey and explain the concept of planetary boundaries (of which more later) to the general public, its message can be made simpler and clearer based on hard science but *socially determined risk tolerances*. Similarly, the *Global Safety Net* has produced a valuable interactive digital map[3] that highlights the important role of indigenous peoples for sustainability but has yet to include their various risk tolerances. To more credibly catalogue the anthropogenic impacts on Earth's biosphere, and hence its sustainability, would require dozens of books. Contrary to the dearth of literature on collective wisdom, literature on sustainability consists of a dizzying array of issues lacking comprehensive synthesising schemas. *Panarchy* as set out in the book by the same name (named after the Greek god Pan) is largely an unrecognised exception. It is an integrative concept of economic, ecological, and institutional systems that seeks to explain unpredictable change across hierarchies and scales. Sustainability has been described as the capacity to create, test, and maintain adaptive capability. Development, however, is the process of creating, testing, and maintaining opportunity [51, p. 76]. Recently, one major

[3] https://www.globalsafetynet.app/

sustainability advance has been recognising Earth as a large, coupled human and natural system consisting of many smaller interlinked systems connected by flows of information, matter, and energy that co-evolve through time as a set of interconnected complex adaptive systems [113]. In the *panarchy* framework, sustainability is accomplished by means of connections among nested sets of adaptive cycles arrayed as a dynamic hierarchy in space and time [114, p. 396]. Initially conceived as an ecological concept, panarchy concepts can be fruitfully applied to the interactions between biophysical and social-ecological subsystems which can then be jointly considered. These subsystems may consist of multiple levels, such as a local watershed within a regional watershed or a set of nested institutions comprised of local and international organisations. Social and ecological subsystems may be linked by either negative (dampening) or positive (reinforcing) feedback loops and are interdependent and co-evolutionary. In a system with multiple nested adaptive cycles, reinforcing feedback may "cascade" downwards to the lowest level or "bubble" upwards to the top. Examples of the latter include environmental actions undertaken by individual citizens that may have cumulative worldwide effects or the use of social media by grassroots demonstrators to influence governments. Horizontal processes within the same panarchy level may also be important, and neither horizontal nor vertical interactions need be sequential as evidenced by pandemics that seem capable of skipping levels [115].

Science has made huge strides in growing our knowledge about how the Earth system functions, of which panarchy is one, but there is increasing anxiety that global decision-making is failing to keep up with the pace of change. We are failing in our new role as planetary stewards, a role that is now crucial to sustaining human development, and even the survival of civilisation itself as we have come to know it [116]. It involves joint collaborative decision-making as the essential mechanism necessary for thoughtful targeted action. As Donella Meadows points out, human mastery over paradigms is the ultimate lever that can be used to change systems [117]. An example looking forward is humanity's foresight to at least set the current seventeen interdependent Sustainable Development Goals (SDG) to be reached by the target date of 2030 as shown in Table 4.1.

The global indicator framework of the SDGs includes 231 unique indicators, which will be reviewed again in 2025 by the United Nations Statistical Commission and can be complemented by indicators at the regional and national levels by member states [119]. In a 2020 study, SDG indicators were compared to a set of external measures which concluded that progress is mostly skewed in favour of socio-economic development as opposed to environmental conservation, and that the SDGs would therefore likely serve as a smokescreen for continued environmental destruction throughout the decade [120]. *The Sustainable Development Goals Report 2024* found that only 17% of the SDG targets are on track, nearly half are showing minimal or moderate progress, and one-third have stalled or even regressed. Against this backdrop governments nevertheless rallied behind the 2030 Agenda with a commitment to rescue the SDGs and to deliver an SDG Stimulus as demonstrated at the SDG Summit in September 2023. In the political

Table 4.1 Titles of the UN Sustainable Development Goals

1	No poverty	**10**	Reducing inequalities
2	Zero hunger	**11**	Sustainable cities and communities
3	Good health and well-being for people	**12**	Responsible consumption and production
4	Quality education	**13**	Climate action
5	Gender equality	**14**	Life below water
6	Clean water and sanitation	**15**	Life on land
7	Affordable and clean energy	**16**	Peace, justice and strong institutions
8	Decent work and economic growth	**17**	Partnerships for the goals
9	Industry, innovation, and infrastructure		

Source: [118]

declaration adopted by the General Assembly, member states noted that "the achievement of the SDGs is in peril" and stated their determination "to make all efforts to implement the 2030 Agenda and achieve the Sustainable Development Goals by the target year of 2030". To get the SDGs back on track, a top priority for the global community should be to rally all stakeholders to end the wars and conflicts that cause unimaginable suffering globally. Sustainable development is impossible without peace. In all countries, reinvigorating just climate transitions is crucial to addressing the triple planetary crisis of climate change, biodiversity loss, and air pollution while reorienting economies towards more sustainable growth. Moreover, renewed commitments to gender equality, as well as significantly increased investments in health, education and social protection are necessary for the dignity of all people of all ages [121, pp. 2,3].

As laudable as establishing the sustainable development goals has been, the goals themselves are not sufficient as they stand since the opposite side of the coin that these pre-determined objectives are meant to address, namely the mitigation of associated global risks, relies primarily on the monitoring of no less than 169 lag indicators of progress at the same time that real-time damage may be occurring at various speeds. What is still needed is a system of risk-driven monitoring and evaluation in near real-time that is as dynamic and adaptable as the phenomena it purports to be measuring. The concept of baselines, lag indicators, and lead indicators in the discipline of monitoring and evaluation needs to be reconfigured so as to provide more actionable information by means of measurement improvements, such as system and sub-system boundary monitoring, "jumping indicators", as well as periodically adjusted thresholds that are linked to tipping point estimations [103]. The global agility, coordination, cooperation, and funding that would be necessary to address these issues is perhaps best suggested by humanity's responses to the recent Ebola, Zika, and COVID-19 public health crises. A standing multilateral ethics body equivalent to those used in human medicine should guide the many operational decisions that still need to be made not only in public health, but all of the sustainable development goals given their interlinked characteristics.

The Common but Differentiated Responsibilities Principle

The strong actually do what they can and the weak suffer what they must—Thucydides

The dual level, joint collaborative decision-making hypothesis that we are unpacking has an analogue in the domain of sustainability in the form of the *common but differentiated responsibilities* principle which was formalised by the United Nations Earth Summit in Rio de Janeiro in 1992. The principle sets out the shared obligation of two or more member states towards the protection of a particular environmental resource while establishing differentiated levels at which individual states can effectively enter into a collective response according to both their capacities and their levels of historical contributions to the problem [122]. The U.N. member states may be thought of as individual actors "collectivised" by networks of international trade and influence. Of all the policy frameworks to gain near universal support recently, none is more compelling than that of sustainability. The concept was put on the international policy agenda in 1987 by the Brundtland Commission which formulated the classic definition of sustainable development as that which "seeks to meet the needs and aspirations of the present without compromising the ability to meet those of the future". The same goal has guided other United Nations policy initiatives, notably the Earth Summit of 1992 in Rio de Janeiro and the climate negotiations that began in Kyoto in 1997, which have continued on through several Conference of the Parties (COP) plenaries. Future policy efforts can build on the renewed attention paid to the issues identified in international fora, including the decades-long accumulated experience with various governance approaches in different countries. Between 1998 and 2010 there were more than 1500 national climate laws and policies worldwide—a five-fold increase—which by 2012 covered 67% of all emissions. Sub-national and some city governments led the way with benefits spilling over to other parts of their respective countries [123, p. 8].

These multi-speed, multi scale changes pose new threats, but also new opportunities for humanity. In paying attention to sustainable development, we must overcome the obstacles posed by a lack of understanding, lack of adaptive capacity, and the unwillingness to change. At the closest intersection of these three challenges also lies the ideal latent solution comprised of the appropriate actions that we would be willing, able, and wise to take [47, pp. 361–362]. Our future Earth demands imaginative breakout events that harness the best collaborations between science and those in society responsible for designing and implementing solutions—industry, civil society, policymakers [116]—and citizens. Research collaboration across a range of disciplines and the coproduction of knowledge among scientists, funders, and stakeholders will be crucial to bridging research and policy barriers to sustainability. Use-inspired primary research that links knowledge to action is the emerging paradigm of sustainability science. Its aim should be to understand the fundamental character of interactions between nature and society by fostering problem-driven, interdisciplinary collaboration, and integrating the effects of system processes across scales from local to global and across different social actors and ways of knowing [124].

The "Greek Tragedy" of the Commons

You can take a horse to water but you can't make him think—Adam Grant

From hunter-gatherer societies to nation-states cooperation is a fundamental organising principle of human societies. We might expect decision groups to fail because everyone has an incentive to free-ride on the work of others, yet both modern and ancient societies have depended on cooperation more so than the rest of the animal kingdom. Cooperation is understood as a contribution to an outcome that benefits all members of a group irrespective of whether each group member has contributed to the outcome. Contributors, also known as cooperators, incur a cost when doing so in contrast to free-riders who do not. Since free-riders gain more than cooperators, they have an evolutionary advantage if the gains can be used to increase their survival and offspring. The inherent tension between individual and group objectives is known by several names, including the collective action, common-resource, or public goods problem. Decisions about cooperation must have had a profound effect on survival and reproduction in our ancestors. Mathematicians who create models of such behaviour have shown that natural selection favours the use of *reputation as a decision rule* of whether or not to cooperate with others. When good or bad stories are transmitted, they influence people's reputations. Several theories have been advanced for the persistence of human cooperation, including direct reciprocity and indirect reciprocity. The basic idea of direct reciprocity is that the potential of cooperation tomorrow can incentivise cooperation today. Indirect reciprocity, however, presumes that most group interactions are observed and can subsequently be shared within the group thereby making *gossip* a powerful selection mechanism [28, pp. 306–310]. Indeed, we are the "storytelling animal" [125].

The tragedy of the commons was recognised as early as Aristotle who noted that "what is common to the greatest number has the least bestowed upon it. Everyone thinks chiefly of his own, hardly at all of the common interest".[4] The expression "tragedy of the commons" has become popular since it was first coined by Garrett Hardin in 1968. He used the example of a cattle herder who in a society that believes in the freedom of the commons increases his herd by overgrazing his animals at a direct cost to his peers, in this case, the loss of preferred pasture. The tragedy of the commons can be readily understood by application of the public goods concept. It is a matter of a specific resource that is *non-excludable*, but which at the same time does not satisfy *non-rivalry*. In other words, one herder cannot exclude another from a common pasture, but once the grass is eaten by one's herd, it is no longer available to another. If the only commons of importance were a few grazing locations or fisheries (the other most popular example), the tragedy of the commons would be of little general interest. Instead, it has been used to describe such diverse problems as firewood crises throughout the less developed world, the problem of acid rain, and the communal conflict in Cyprus [126, pp. 2–3], among others. Both

[4] *Politics*, Book II, ch. 3.

public goods and the tragedy of the commons belong to a broader class of conditions known as *collective action problems*. The strategy of first pooling and then averaging out costs and benefits has direct relevance to such problems. An important case is Tradeable Environmental Allowances that set a limit on environmental withdrawals or emissions and permit their free trade in market-based systems based on allocated allowances [56]. This was one of the bases of the Kyoto protocol, and more recently, carbon budgets and carbon trading schemes to combat climate change. Such methods *dissipate global risks* (not necessarily the thing itself in absolute terms) thereby helping to make the human-environment system semi-resilient rather than resilient, but also more equitable.

Collective action problems tend to be easier to solve the smaller and more homogeneous the actors involved and the better the availability of information. Families generally solve collective action problems effectively; international organisations find it much harder to do so. Global efforts to reduce carbon emissions require coordination among a large group of diverse actors who possess only imprecise monitoring mechanisms. Solving such a complex problem requires coordination as well as an enforcement mechanism. History has taught us that overgrazing meadows and overfishing waters invites the risk of collapse. The same precautionary tale applies to the collective action problems that we face today. The late Elinor Ostrom spent decades studying real-world efforts to address collective action problems. She found that in addition to monitoring deviations, communities that solve collective action problems agree on clear boundaries, agree on well-defined rules, grant the authority to impose graduated sanctions, and establish mechanisms to resolve disputes [126, p. 281].

In the context of limited resources, general acceptance of the type of commons ownership—government, private, or community—helps but is not uniformly successful in stopping major environmental deterioration as shown for forests in multiple countries. Sound science is necessary but not sufficient for commons governance. Too many strategies for local commons governance are designed in capital cities and by donor agencies unaware of the state of the science itself and local conditions. The results are often tragic but localised. But as the human footprint on the Earth grows, humanity is challenged to develop and deploy understanding of large-scale commons governance in time to avoid large-scale tragedies that will otherwise ensue [56]. It is now time to create a new category of institution, the "Commons Sector", which would be responsible for managing existing common assets and for creating new ones. Some assets should be held in common because it is more ethical, others because it is more efficient; and yet others because it is more sustainable. Common Asset Trusts have been proposed for managing the commons sector at various scales. Trusts can *propertise* the commons but without privatising them. The Alaska Permanent Fund is one example, along with the many land trusts currently in existence. Common asset trusts could protect or restore critical natural capital—those resources provided by nature, such as bee pollination, that are in some way essential to human well-being [127].

The Anthropocene

Man has introduced into the planet's structure a new form of effect upon the exchange of atoms between living matter and inert matter—Vladimir Vernadsky

Following guidance from the Subcommission on Quaternary Stratigraphy and the International Commission on Stratigraphy, the Anthropocene Working Group resolved in a binding vote in May 2019 to formalise the *Anthropocene*. It would be defined by the standard means for a unit of the Geological Time Scale, via a Global boundary Stratotype Section and Point [128]. The formal establishment of an Anthropocene Epoch would mark a fundamental change in the relationship between humans and the Earth system. Wider recognition that human actions are driving far-reaching changes to the life-supporting systems of Earth may well have increasing philosophical, social, economic and political implications over the coming decades [129]. The knowledge that human activity now rivals geological forces in influencing the pathway of the Earth System as a whole has important implications for both the Earth sciences and global decision-making [130]. Recent global environmental changes indicate that Earth may have entered a new human-dominated geological epoch—the Anthropocene. The epoch and term are currently not a formally defined geological unit within the Geological Time Scale as intended when coined by Paul Crutzen and Eugene Stoermer in 2000. While its value as a formal geological phenomenon is being deliberated, the current informal use of the term has already proven highly useful to the global change and Earth System research communities, as well as a popular scientific term. The case for a new geological epoch seems reasonable. Human activity has now become global and the dominant influence of most contemporary environmental change [129, 131].

A few million years before *Homo sapiens*, the genus *Homo erectus* had already mastered the *techne* of making stone tools and rudimentary weapons. They subsequently learned to control fire, a crucial breakthrough that fundamentally altered our relationship with other animals, none of which could manipulate fire. Control of fire undoubtedly helped hominids in their hunt for food and to keep dangerous animals away from their camps at night. The increasing access to a protein-rich food source had other benefits for early humans. The change from a primarily vegetarian diet to an omnivorous diet triggered a fundamental shift in the physical and mental capabilities of early humans, the latter arguably being the more important. Brain size grew threefold to about 1300 cm^3 giving humans the largest brain-to-body ratio of any animal on Earth. This allowed the development of spoken language first, and then written language, both of which were instrumental in the accumulation of knowledge from generation to generation. Yet our omnivorous ancestors could only extract from plants what energy they had captured for photosynthesis which amounted to 1% of incoming solar energy, while animals eating plants could extract only about 10% of their stored energy. These energy constraints limited the growth of human numbers and activity until the discovery and use of fossil fuels. Fossil fuels represented a vast energy store of solar energy from the past tens, or hundreds of millions of years of photosynthesis. They were the perfect fuel source—energy-rich, dense, easily transportable, and relatively easy to access. Human energy use multiplied. The energy use of our hunting and gathering forebears multiplied three or four times up until our agrarian predecessors, whose energy use multiplied again four or five times until the development of industrial societies. Ultimately this progress led to an explosive store of knowledge upon which humanity built complex cultures and societies, including the power to affect and manipulate the environment [131].

Human activity has profoundly affected the environment, from the evolution of life to Earth's major biogeochemical cycles. Notably, the early twentieth-century invention of the Haber–Bosch process, which converts atmospheric nitrogen to ammonia for use as fertiliser, has altered the global nitrogen cycle so fundamentally that the closest suggested geological comparison concerns events from about 2.5 billion years ago. Human actions have released 1 billion metric tons of carbon to the atmosphere since 1750, increasing atmospheric CO_2 to a level not seen for at least 800,000 years, and possibly longer, thereby delaying Earth's next glaciation event. The added carbon has increased ocean water acidity at a rate probably not exceeded in the last 300 million years [129]. Every indicator of human activity underwent a sharp rate of increase around 1950, all from very low or non-existent bases. The change was so dramatic that the period from 1945 to the present has been called the Great Acceleration. For example, the consumption of petroleum grew by a factor of 3.5 since 1960, and motor vehicles rose from 40 million at the end of World War II to about 700 million by 1996 and continue to rise steadily to this day [131]. Other phenomena associated with the Anthropocene include:

- The proliferation and global dispersion of many new "minerals" and "rocks" including concrete, fly ash, plastics, and the myriad "technofossils" produced from these and other materials.
- Distinct anthropogenic perturbations of the carbon, nitrogen, and phosphorus cycles, as well as perturbations from various metals and new chemical compounds.
- Environmental changes generated by these perturbations, including global warming, sea-level rise, ocean acidification, and the spread of oceanic "dead zones".
- An order-of-magnitude increase in erosion and sediment transport due to urbanisation and agriculture.
- Rapid changes in the biosphere both on land and in the sea as a result of habitat loss, predation, and an explosion of domestic animal populations and species invasions.

Many of these changes are altering the trajectory of the Earth System and will persist for millennia or longer, some with permanent effect. They are accumulating in distinct geological strata colloquially known as "golden spikes". The clearest and most globally synchronous of these phenomena was produced by the artificial radionuclides spread worldwide by the thermonuclear bomb tests of the early 1950s, which can now serve to mark the beginning of the Anthropocene epoch. A recent study has shown that human-made mass surpassed the overall living biomass of the

Earth in 2020 (±6 years). At the beginning of the twentieth century, anthropogenic mass was about 3% of global biomass on a dry-weight basis. One hundred and twenty years later, anthropogenic mass exceeds overall biomass in the world: 1108 gigatonnes (buildings, infrastructure, and plastic) versus 904 gigatonnes (trees, shrubs, and animals). Over the past 100 years, anthropogenic mass has increased rapidly doubling approximately every 20 years in contrast to total biomass which has remained reasonably stable. This amounts to each person on the globe producing more than his or her body weight in mass every week [132] and an average social metabolic rate in the United States of 11,000 watts compared to the historical 90 watts needed per day to maintain life [31, p. 13]. This story though began millions of years ago.

The term *Anthroposphere* is a complementary concept to the Anthropocene and may be thought of as the addition of a spatial layer to the Anthropocene's temporal dimension. Three scientific journals focusing on the topic have been launched: *The Anthropocene, The Anthropocene Review,* and *Elementa.* A key attribute of the Anthroposphere over the past century has been a rapid world population growth with estimates that between 8 billion and 10.5 billion people will be living on Earth by 2050. Human dwellings now occupy about 8% of ice-free land with about three-quarters of the land surface already having been altered by humans in many ways and scales. Humanity's appropriation of fossil fuel energy, and the many technologies it powers, has played a major role in amplifying the impact of the Anthroposphere on the Earth system. It has enabled humans to extract resources such as iron ore and bauxite, which are used to make automobiles, skyscrapers, and numerous other artefacts integral to modern life. A defining decision for humans this century will be to either follow the current unsustainable course or to alter the amount and nature of human activities so as to improve the interactions among the anthroposphere, geosphere, biosphere, atmosphere, and cryosphere—all within the boundaries of the Anthropocene [133].

Safe and Just Planetary Boundaries

For the first time, we have the power to decide the fate of our planet and ourselves—
Carl Sagan

In 2009 Johan Rockström and his collaborators proposed the concept of *planetary boundaries* for the purpose of estimating a safe-operating space for human sustainability and development with respect to the functioning of the Earth System. This initiative was welcome because the existing paradigm of social and economic development based on economic growth remained largely oblivious to the risk of human-induced environmental disasters at planetary and continental scales. Nine interlinked planetary boundaries were identified, seven of which were scientifically quantified. Planetary boundaries in the proposed framework were to be *socially determined* so as to be at a safe distance from the nine corresponding *biophysical thresholds* that determine the self-regulating capacity of the Earth system. The transgression of a threshold could trigger non-linear changes threatening the conditions which have enabled humanity to evolve and thrive, namely, those that prevailed in the geological epoch of the Holocene prior to the industrial revolution [134]. The planetary boundaries framework provides a science-based approach to establish that human forcing has already, at the planetary scale, rapidly pushed the Earth system away from Holocene-like conditions, and onto an accelerating Anthropocene-like trajectory. In recent years, there have been multiple additional efforts to deepen the understanding of planetary boundaries and the safe-operating space for humanity. Attempts to quantify interactions between planetary boundaries suggest that cascades and feedbacks mostly amplify human impacts on the Earth system and thereby shrink the safe-operating space for humanity.

In the 2015 update of the study the authors estimated that two planetary boundaries and two thresholds have already been transgressed. Respectively: climate change, land-system change, genetic diversity, and biogeochemical flows. The original 2009 study was more suited to governance issues compared to the 2015 version which downplayed the need for socially determined boundaries thereby reducing possible societal responses to being reactive as opposed to proactive [135]. However, the 2023 follow-up version of the paper reversed this trend. At least three notable improvements were made. First, it added a "just" component (interspecies, intergenerational, and intragenerational). Second, it was made more consistent with IPCC methodologies by providing confidence levels for the measured variables. And third, by de-emphasising the previous use of the term "thresholds" in favour of the more commonly understood "boundaries", it was made clearer for public consumption. In addition to the necessary data update and review (online supplementary data, methodology details, and coding), the authors conclude that seven of the eight global-scale safe and just Earth System Boundaries (ESBs) have already been crossed. The effects are spatially widespread, with two or more ESBs transgressed for 52% of the world's land surface, impacting 86% of the global population. Some communities experience more transgressions, with four or more ESBs exceeded for 28% of the global population, but only 5% of global land surface. Spatial hotspots are concentrated in regions of higher population density, raising major intragenerational justice concerns. These are a part of the so-called three "I s" which include interspecies justice and intergenerational justice. The authors assert that an assessment of safe and just boundaries is required that accounts for Earth system resilience, as well as human well-being in an integrated framework. The selected domains are the climate, biosphere, freshwater, nutrients, and air pollution because they span the main components of the Earth system, namely, the atmosphere, hydrosphere, geosphere, biosphere, and cryosphere by interlinked processes of carbon, water, and nutrient cycles [136].

The most recent 2023 update of the planetary boundaries framework calculated that six of the nine boundaries have already been transgressed, namely, *climate change, change in biosphere integrity, land-system change, freshwater change, biochemical flows (phosphorus and nitrogen),* and *novel entities*. Conversely, *ocean acidification, atmospheric aerosol loading,* and *stratospheric ozone depletion* were deemed to be still in the safe-operating zone. This update also introduced *net primary production* as a measure of functional biosphere integrity by quantifying the energy available to ecosystems versus the energy available to human societies. Additionally, the "zone of increasing risk" was introduced to replace the "zone of uncertainty" as the indicator for transgressed boundaries rather than the "zone of uncertainty" since the demarcation of this zone was not just based on plain scientific uncertainty. A large body of recent research supports the conclusion that the climate change and biosphere integrity boundaries are in a zone of rapidly increasing and systemic linked risks. This strengthens the use of the precautionary principle to set the planetary boundaries at the lower end of the zone of increasing risk [137]. This is a welcome change to the framework that better links science and risk, both logistically and semantically. Paradoxically, the planetary framework was in need of such subjectivity if risk management was to be taken seriously given that risk perceptions are a key input of comprehensive risk assessments. Multi-criteria decision analysis (MCDA) is a useful technique in this regard.

The behaviour of many systems is largely determined by decisions made by individuals and groups within those systems. In population systems, for example, the behaviour of couples of child-bearing age determine the dynamics of the system; in market systems, the aggregate decisions of consumers drive market behaviour; in industries, such as the electricity sector, decisions by corporate executives on the generation mix determine many characteristics of that system. In consequence, to understand the behaviour of systems, understanding the nature of decision-making within those systems is critical [138]. MCDA methods have become increasingly popular in large part due to the multifaceted challenges involved with our global sustainability goal and the complexity of socio-economic and biophysical systems. Multi-criteria analysis consists of four main stages: *criteria selection, criteria weighting, operational evaluation,* and *final results aggregation.* Criteria selection primarily involves economic, technical, environmental, and social factors. Weighting criteria may be subjective, objective, or a combination of both. Aggregation methods fall into two broad categories: voting methods and mathematical aggregation. As an operational evaluation approach, MCDA is suitable for addressing complex problems featuring conflicting objectives, different forms of data and information, multiple interests and perspectives, and dynamic socio-physical systems [139]. A common metric used to address risk is the composite risk index, which is defined as the impact of the risk event multiplied by the probability of its occurrence. The impact is usually expressed as the dollar cost of the estimated damage, and the probability by some version of a power law (one quantity varies as a power exponent of another) [54, p. 315]. Additional criteria such as "urgency' and "intractability" can be included to create a more comprehensive risk profile. Non-subjective weighting and computational aggregation can complement qualitative approaches but can get highly technical and perhaps better left for experts to facilitate final decision-taking.[5] As society becomes ever more complex and risk averse, further developing a

[5] **Subjective Weighting**: PATTERN consistent matrix analysis; simple multi-attribute rating; Swing; pair-wise comparison; analytic hierarchy process; least-square method; trade-off; eigenvector method; Delphi method; SIMOS; SMARTER. **Objective Weighting:** least mean square method; min-max deviation method; entropy method; TOPSIS method; vertical and horizontal

science of risk is becoming increasingly important [54, p. 315]. More specifically, criteria selection is the most important stage and is amenable to participatory processes, such as Delphi, town halls, referenda, and other democratic and combinatory approaches.[6]

But to be effectively executed as originally intended, the ESB framework would require the equivalent of about eight more Paris Agreements. Devolution to nation-states—currently the key locus of implementation in the international system—would require repeated national consensus so as to reset precautionary boundaries ahead of approaching thresholds that are transnational and subject to extensive ongoing quantification. It is likely that future measurement will face similar challenges to the ones that the SDGs currently face, especially if the ESBs do not make it sufficiently clear that quantification at given *time intervals* of the upper and lower bounds of all the control variables is what should guide public policy. The prime example is the Paris Agreement by 189 countries to progressively limit greenhouse gas emissions by the recommended date of 2030 so as to keep global warming under 1.5 ° C relative to pre-industrial levels in order to avoid the threshold of 2^0 C by 2050 when abrupt changes to the Earth system are expected to occur. It is also essential to emphasise that the ESBs rely primarily on objective, independent measurement, whereas the SDGs are largely reliant on self-reporting by nation-states. Beyond the issue of measurement, the overlapping objectives between the two frameworks (SDGs and ESBs) reduce attention unnecessarily on one or the other. Ways of folding-in the ESBs into the SDGs is one way of addressing this problem, for example, by incorporating additional objectively verifiable targets and specifying so in footnotes or other means. Seriously addressing these issues, together with any related emergent ones, will make the public appeal of the boundaries framework more impactful for governance purposes and sustainability.

Societal Collapse and Recalibrating the Clock of Sustainability

If we are to survive, our loyalties must be broadened further, to include the whole human community, the entire planet Earth—Carl Seagan

Failure of wise decision-making at the highest group levels can lead to societal collapse. Jared Diamond in his book *Collapse* suggests four ways in which such failures have occurred historically and may reoccur now. First, a group may fail to

method; variation coefficient; multi-objective optimization; multiple correlation coefficient; principal component analysis. **Combination Weighting**: multiplication synthesis; additive synthesis; optimal weighting (sum of squares); optimal weighting (minimum bias); optimal weighting (relational coefficient of gradation).

[6] See Review on Multi-Criteria Decision Analysis Aid in Sustainable Energy Decision-Making by Jiang-Jiang Wang et al. and A Review of Multi Criteria Decision Making (MCDM) Towards Sustainable Renewable Energy Development by Abhishek Kumar et al.

anticipate a problem before it arrives. For example, the deliberate introduction of foxes and rabbits by British colonists into Australia has meant that foxes decimate many native species, while rabbits outcompete many local herbivorous mammals with a cost in the billions of dollars to control. Second, when the problem does arrive, the group may fail to perceive it as such. This is exemplified by climate change and "creeping normalcy" which is characterised by slow trends concealed in noisy fluctuations. Third, after the group perceives a problem, they may fail to address it in part due to a clash of interests one of which is "the tragedy of the commons". Central to this failure is the "free-rider" problem whereby an incentive exists for immediate resource extraction of a limited resource before a competitor can do the same, but to the eventual detriment of all. A case in point is the collapse of most of the world's major marine fisheries. Finally, a group may try and solve a problem, but not successfully. For example, although forest fires could be greatly mitigated by thinning the understory of a forest and removing fallen timber, the cost of doing so on a large scale is prohibitively expensive [140, pp. 420–437]. And matters can get even worse by the interaction of some of these factors as tragically displayed in California in 2020 (and after) by record breaking forest fires exacerbated by climate change. Throw in human foibles like corruption into the mix of any hazardous situation and watch it become grave.

According to J. Tainter, societal collapse is an abrupt, pronounced loss of an established level of socio-political complexity. Such a collapse results in a society that is suddenly smaller, less complex, and less socially differentiated. There is less centralised control while specialisation decreases. The flow of information is reduced, people trade and interact less, and there is lower coordination among individuals and groups overall. Economic activity drops accordingly, while the arts and literature experience such a quantitative decline that a dark age often ensues. Population levels may drop, and for those who are left, the known world shrinks. This is because decoupling economic growth from energy and material inputs runs counter to physical laws: basic needs for food, shelter, employment, and education for the billions of people that will be added by 2100 will alone demand substantial energy flows and material inputs. Although those inputs will have lower relative intensities than today's average rates—the absolute totals will keep rising with continuing population growth or will moderate but remain substantial. This has been modelled by J. D. Ward and colleagues such that they concluded that GDP growth cannot be sustained indefinitely. In consequence, it makes it highly misleading (if not irresponsible) to advocate for any growth-oriented public policies assuming that a decoupling and continued GDP growth are even possible [59, p. 492].

To avert societal collapse, according to Vaclav Smil, the following human agency is needed for sustainability:

> On the global level, there is no greater what-comes-after-growth uncertainty than the very fate of modern high-energy civilisation with its still continuing population growth, high material requirements, and commensurately high environmental impacts. All of these long-lasting trends will have to end, deliberately or involuntarily. There is no possibility that we will be "saved" by an early coming of the Singularity or by an early terraforming of Mars. Such fictions make great news headlines but are worthless for dealing with civilisational

> challenges. Dealing with them is further complicated by continued profound uncertainties as we still have only approximate ideas about tolerable biospheric boundaries and hence of the timing of deliberate epochal shifts onto a new existential plane…*Wrong decisions can accelerate decline and demise,* while proceeding cautiously could greatly limit most (if not all) undesirable biospheric and social outcomes [emphasis supplied] [59, pp. 454, 479].

In a 2021 review article, Carl Folke and his co-authors sound similar warnings and propose the following four major pathways towards global sustainability: acknowledge and act on the evidence that societal development is entangled in, and critically dependent on, the biosphere and the broader Earth system; create incentives and design policies that enable societies to collaborate in pathways of sustainable and just futures within planetary safe zones; enhance the resilience of the biosphere by creating transformational stewardship of social, economic, and cultural development; and make active use of converging and emerging technologies to facilitate the transformation of societal stewardship [99] (Fig. 4.1).

Supporting and enhancing learning strategies and capacities for dealing with shocks and stressors so as to deal with change and development in the turbulent times of the Anthropocene will be instrumental to combatting inequity and improving the quality of life. Multiple inequities will require multiple sustainabilities and diverse forms of responses attuned to diverse contexts yet all directed towards transformations for Earth's sustainability [99]. Attaining the current seventeen Sustainable Development Goals also requires publicly cogent sustainability metrics—without a *sustainability clock* for citizens to get behind as evocative as the nuclear doomsday clock we encountered earlier—scientists and politicians will lack the necessary

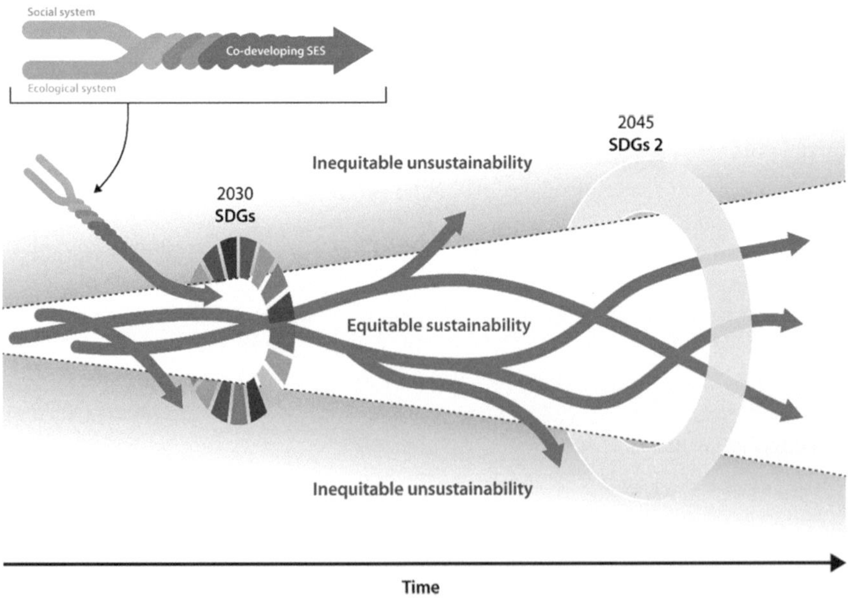

Fig. 4.1 The SDGs cone of equitable sustainability. (Source: Adapted by [99] from [141])

vision and will to think and act as if their own well-being was at risk. Past human societies did not have either the foresight or the means by which to gauge the unintended consequences of their collective actions, but current societies have that capability. The most visible undertakings in this regard have been the negotiations of world leaders under the aegis of the United Nations in Copenhagen in 2010, and more productively, in Paris in 2015 so as to discuss and agree on treaties for climate change. The conferences, known as COPs and sequentially numbered, had to distil the views of over 200 countries all of which had to align and integrate economics, politics, history, and ecology. They used many of the techniques developed by the craft of diplomacy, including "talks about talks", as well as other substantial preparations. And importantly, they also drew on one of the greatest exercises in orchestrated science: the Intergovernmental Panel on Climate Change (IPCC). The IPCC is an extraordinary assemblage of knowledge and intelligence involving the pooling of data, construction of detailed scenarios of possible futures, modelling the climate by powerful supercomputers, and soliciting and mobilising thousands of scientists to provide inputs.

Similarly, the C40 group of cities, the Copenhagen Consensus follow-up conferences, the Climate CoLab, and corporate inputs on the measurement of carbon emissions have made their own significant contributions to the state of knowledge. All these developments have occurred in the context of watchful NGOs, as well as expert commentators and groups that synergistically constitute a global information commons that transforms knowledge about a global threat into a shared public good in need of a solution [24, p. 64]. Top-down governance proposals such as a World Environment Organisation or an Intergovernmental Panel on Global Sustainability could contribute solutions, but nonetheless would require bottom-up citizen inputs to ensure their long-term viability and short-term acceptability [98, 142]. Widespread motivation already exists to enhance institutional mechanisms that enable citizens in a democracy to make comparatively better decisions while sustaining core values, and in consequence, to advance common societal interests [143, 144].

Chapter 5
Dynamic Epistemic Equilibria

We believe it right to look for what we don't know than if we believe there is no point in looking because what we don't know we can never discover—Socrates

The proposed theory of *dynamic epistemic equilibria* that follows begins with our biology. Summarising recent cutting-edge neurobiological research, Joseph LeDoux reports that according to higher-order theory (HOT), conscious awareness comes about when nonconscious first-order sensory information is cognitively *re-represented* resulting in a higher-order state, which is in essence a mental state about a lower-order mental state. First-order theory focuses on the mental state that represents the world, while HOT adds another higher mental state that re-represents the original sensory state. Put more simply, the second higher-order state makes conscious the first-order state. To become conscious of the content of the second higher-order state itself, however, requires that it in turn be re-represented by an extra higher-order, third state. This version of HOT is known as the "higher-order representation of a representation theory" or HOROR theory. These human characteristics have been traced down to the neuron level at the Frontal Pole of the prefrontal cortex as the most likely area for high-level conceptual processing, which is generally considered to have the greatest capacity of any brain area for this type of function. The Frontal Pole is considered to be more involved in representing long-term goals, as well as cognitive multitasking and hierarchical reasoning. This capacity to fashion a mental "big picture" would seem to be a key feature and advantage of humans. It follows that in the absence of identified and agreed upon mechanisms that might work similarly to HOROR at the group or social level, a theory of higher-order wisdom is needed to shape existing knowledge into an explanatory, coherent whole down to the individual level [25, pp. 278–283]. A distinctive feature of human cognition is the capacity to use concepts to engage in *hierarchical relational reasoning* which enables us to simultaneously consider multiple options—to multitask when engaged in deliberation and problem-solving. Critics of higher-order theory argue that prefrontal areas of the (individual) brain lack the types of fine-grained representations that can support the rich quality of human perceptual experiences. However, this critique loses much of its relevance when perceptual consciousness is viewed not merely as a sensory-dictated perceptual experience, but as a top-down construct based on combinations of sensory, memory, and conceptual

E. Bellos, *Sustainability Thinking*, https://doi.org/10.1007/978-3-032-11244-6_5

representations that enable rich perception via conception. Joseph LeDoux summarises these ideas as follows:

> The pattern of inputs to the higher order network is compatible with the multistate hierarchical higher-order model…in which the network can draw upon a wide range of states, from pure sensory to mnemonic and conceptual, and many in between, in constructing schema/mental models that underlie our conscious experiences [25, pp. 232–233, 308–309].

As the philosophical physicist David Bohm has noted, the whole ecological problem is due to thought because we have thought that the world is there for us to exploit, that it is not finite, and so no matter what we did the pollution would all get dissolved away. Thought has to be in some sense aware of its consequences, but currently thought is not sufficiently aware of its consequences. We could say therefore that practically all the problems of humanity are due to the fact that thought is not as *proprioceptive* (the body's awareness of its position in space) as the body is. More simply, we could speak of the "self-perception of thought" or "thought awareness of itself in action"; that is, the process of thought in which there should be awareness of its movement, of the intention to think, and of the result that that thinking produces [43, pp. 11, 28, 91]. Along these lines, we can proceed in search of a *collective meta-awareness seeded by sustainability thinking and a reverence for all life*. It is particularly difficult to converge on this objective from either the bottom or the top. From the bottom, when consciousness is considered as mind (as opposed to consciousness as waking state, or experience), we have to take into account that less than five percent of brain activity is conscious; and when we reason, solve problems, make judgements, or take decisions, our thinking is, moreover, subject to the primacy of affect[1] [5, pp. 184, 187]. From the top, a change of orientation is necessary from the currently dominant focus on the individual since there is no collective brain, only collective processes and inheritances such as *biophilia* (the love of nature) that inhere in preparation for desired outcomes in what may amount to an ethereal collective mind. From a physics perspective, "quantum theory implies that the universe is basically an indivisible whole, even though on the larger scale level it may be represented approximately as divisible into separate existing parts. In particular, this means that, at the quantum theoretical level of accuracy, the observing instrument and the observed object participate in each other in an irreducible way. At this level perception and action therefore cannot be separated".[2] Furthermore, most of our thought in its general form is not individual, it originates in the whole culture and is pervasive. This constant flow creates a deep structure of thought, which is what is common and subject only to being continuously drawn down or replenished. The content and deep structure of thought are not really separate because the way we think about thought has an effect on its structure, in consequence both structure and content have to be looked at. All of

[1] "Affect" may be too readily equated with emotion, although emotions certainly are a part of it, the implication is much broader: it is a way of attending to the world, a disposition towards the world (McGilchrist, I., 2019, p. 184).

[2] Quoted in [145, p. 222].

this thought, which is based on what one has been thinking, stems from memory. Knowledge is built up through experience and practice. You think about it, organise it, and it goes into memory as knowledge. Part of that knowledge is acquired skill through practice, that too is a kind of memory stored somewhere in the body or in the brain, it is all part of one system [43, pp. 59–60].

Against this background, the proposed theory of dynamic epistemic equilibria is analogous to being both the collective outcome and the collective process leading to that very same outcome—similar to one's individual judgement being both the mental activity of making a judgement as well as the result of that judgement [82]. It is made up of unified collective wisdom-making and wisdom-taking. Dynamic epistemic equilibria work across scales when they can be effectively applied by decision-making groups engaged in epistemic processes derived from the co-mingling of data, information, knowledge, intelligence, imagination, and wisdom as illustrated in Fig. 5.1. The proposed theory draws on the structure and function of epistemic communities, broadly defined, to serve the purpose of collectivising *episteme, techne, phronesis,* and *sophia* (see Fig. 5.3). Also consider that Aristotle's *sophia* is not commonly associated as being inclusive of collective wisdom, nor has the collective process leading to that wisdom been clearly identified. The need for unification of division with union is an important principle in all areas of life; it reflects the need not just for two opposing principles, but for their opposition and asymmetry ultimately to be harmonised in favour of union [5, p. 200]. Ways of organising thought on a large-scale are nascent in part because they lack a generally accepted guiding theory and the necessary expertise for their further development. But new

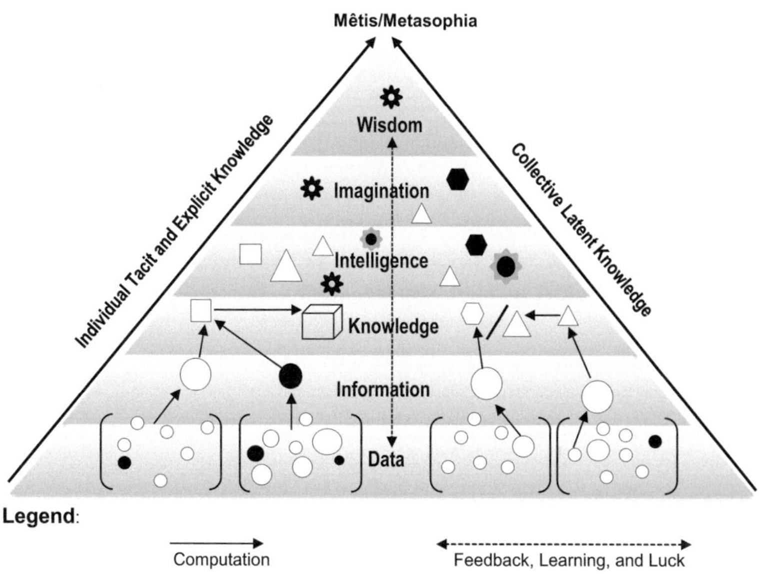

Fig. 5.1 Metasophia's emergence from collective mental metabolism

ways of organising thought suggest that despite a lack of a reliable economic base, the future of almost every field of human endeavour could become better at harnessing information and learning quickly [24, p. 10]. In scientific pluralism, the ultimate goal of scientific inquiry is not to arrive at a single theory necessarily, but rather to fill the gap where no unified theories are available to explain a phenomenon, or where the phenomenon can only be explained by multiple theories [146]. Both in coarse and fine-grained modes, the proposed theory of dynamic epistemic equilibria can help organise thinking about one of the most intractable challenges in human affairs—the process and results of *wisdom-making and wisdom-taking with a purpose.* The dichotomy between decision-making and decision-taking plays out most significantly for global sustainability among international bodies and the member states that are expected to implement relevant agreements, but which may do so inconsistently or incompletely and without recourse to clear remedies.

Steven Sloman and Philip Fernbach in their book *The Knowledge Illusion* assert that individual knowledge is remarkably shallow. Human intelligence resides not in individual brains but in the collective mind. To function, individuals depend not only on accumulated knowledge in their memory, but also knowledge stored in our bodies, in the environment, and in other people. In consequence, intelligence and wisdom[3] are a product of a community, not any individual alone [147, pp. 4–5]. Human culture is the culmination and expression of humanity's collective history, while individuals additionally carry their own particular memories. We exhibit cultural variation precisely because we evolved to have this capacity to begin with. Further, each one of us carries within himself or herself the evolutionary blueprint for making a good society. This "social suite" is comprised of individual identity, love, friendship, social networks, cooperation, in-group bias, mild hierarchy, plus teaching and social learning [148, pp. xxi, 13, 14, 17]. Incorporate data, information, knowledge, intelligence, imagination, and wisdom as "metabolites", then mêtis and metasophia become visible as illustrated in Fig. 5.1. The study of wisdom in different eras and civilisations has confirmed that there have been surprisingly convergent views of what counts as wisdom. These include the ability and willingness to take a long view, attention to the details of context as opposed to the mechanical application of rules or heuristics, and the integration of ethics into thought and decisions. Wisdom has generally been taken to mean avoidance of rigidity of mind, fixed methods, and ways of thinking that are not conducive to the nature of the moment and problem. Today we recognise wisdom in leaders who are cognizant that their organisation is part of a sector and economy, their military is part of a larger system that may preserve peace, their city is part of the nation and the world, and that themselves and their community depend on the biosphere [24, p. 224].

From the collective perspective, for the first time in history we have the technological means to function as a global village of nations, but this requires establishing and scaling-up universal mechanisms that make cooperation possible across all

[3] Although it appears in the sub-title of the book, wisdom is not a part of Sloman's and Fernbach's original formulation in this context.

levels. The challenges of scaling-up might seem insurmountable, but a good first step is to establish the theoretical necessity for doing so [29]. The theory of dynamic epistemic equilibria is founded not only in received theory and philosophy, it is also aligned in its unity of process and outcome with modern evolutionary theory in the form of *teleonomy,* which is conceived as both the product and cause of natural selection, not just its mechanism [74]. Moreover, modern computing has not only enabled untold access to information, but it has also forced us to rethink the very nature of knowledge, and so of epistemology, and so of the human mind—the entity which knows—and the philosophical study of that entity, the philosophy of mind. The new sciences of mind, evolutionary psychology, cognitive neuroscience, and others have separately and jointly increased the explanatory powers of how the mind works to the point that both ethics and philosophy of mind have fallen into the sight of science, including the sight of functional magnetic resonance imaging [57, p. 37]. Just as cognitive science has drawn on many sources to understand how people think, so will a new discipline concerned with collective thought need to draw on many sources—from neuroscience to computer science, from social psychology to sociology, from economics to ecological economics—and use these to guide practical experiments. As the new discipline emerges aided by neighbouring disciplines in the spirit of a commons, it will need to become useful by guiding and being guided by a community of practitioners working to design, as well as operate tools that help systems think and act more successfully [24, p. 5].

Dynamic Epistemic Equilibria as Coarse-Grained Theory

> *The world is not divided into two types of being, one superior and the other merely surrounding it. Being, nature, the universe—they are all one infinitely complex and mysterious metaorganism of which we are a part, though a unique one*—Václav Havel

The gap between transnational rule-makers and individual rule-takers has been referred to as the global democratic deficit and addressed in the literatures by two schools of thought, one emphasising structures, the other processes. The former advances idealised models of global democracy, while the latter advocates the values of democratisation [149]. Neither approach, however, is wholly consistent with this book's normative proposal for an *open epistemic democracy* driven by sustainability thinking. Democratic institutions, where some of our most consequential decisions are made, have proven themselves to be incapable of learning how to learn, frozen in structures that are now anachronisms. Many democratic institutions—parliaments, congresses, and parties—appear to be less intelligent than the societies they purport to serve making it easier for the enemies of collective intelligence to capture public discourse, spread misinformation, and fill debates with distractions rather than facts. Even if electoral systems worked better, representatives can still be corrupt, ignorant, or misguided. The use of methods for deciding on policies might ignore most of what is known about how to decide wisely, while public decision-making can be beset by bureaucratic capture, rent seeking, and logrolling, among other shortcomings. Yet the practice of democracy has proved to be resilient avoiding most of these defects, in part because the formal processes of decision-making have been insulated by complimentary systems for thinking. Democracy has aligned itself with the free media, the sciences, and civil society, while the much-maligned bureaucracy can, most of the time, contain the wilder ideas of politicians and parties. In consequence, democratic forms of discussion have allowed for healthy contention, argument, the airing of alternative views, and decision-making that has a greater chance of success [24, pp. 4, 181,184].

Adopting the dynamic epistemic equilibria theory would dampen the debate between epistemic democracy theorists and their procedural democracy counterparts [146]. Even David Estlund's *epistemic proceduralism* with its heavy emphasis on process cannot get away from simultaneously addressing outcomes since it is basically a conservative adaptation of the principle of procedural fairness applied to cases of morally evaluable outcomes. Note that it is the *first* of existing moral power that grants the authority to require or forbid the action of citizens. A theory that does not recommend any particular actions by any individual does not show that it does not counsel action—it is aspirational—it sets sound standards that are not met, but could be met, and it tells us to meet them. Despite the "epistemic" in its name, the reasons epistemic proceduralism gives the citizen about an outcome are *moral reasons to comply, not epistemic reasons to believe* [150, pp. 10, 106, 108, 267]. It is precisely because morality is a *first* that the average human mind cannot be changed easily, and when collectivised, dynamic equilibria are disturbed if one group's perceived morality is asserted to be superior to another's. In its contribution of a moral dimension and the action that may flow from it in healthy societies, epistemic proceduralism may be thought of as a normative theory of action. What is still needed is a *normative theory of collective action.* Dynamic epistemic equilibria as a theory of action bridges this gap by virtue of its claim that sustainability constitutes the highest moral ground. It asks citizens (ordinary and expert) to (i) comply, (ii) believe, *and* (iii) act. To undertake all three actions involves our evolutionary derived *shared intentionality*. The context of these actions may be in public or private organisations making up regional political entities, which in turn comprise national jurisdictions, and which in their turn form the family of nations with many other possible intermediate permutations, or what Geoff Mulgan calls "assemblies" and Hélène Landemore "open mini-publics".

In the simplest terms, the operational proposition advanced here is that the explicit purpose of democracy ought to be sustainable life. In all its permutations democracy is practically hollow if it exists solely for its own sake. Thus, the proposed first fundamental law of democracy can be stated as: *the goal of democracy is to sustain life*. It follows that this first canon ought to be accompanied by the means to attain it. It is argued further that *open epistemic democracy* is the preferred means of meeting the first law of democracy. This formulation combines the concepts of "open democracy" and "epistemic democracy". The term "epistemic democracy" has been coined to capture the notion of democracy as a way of generating knowledge and tapping into the collective wisdom of citizens. It is a reinterpretation of democracy in which large publics take on systemic problems and make

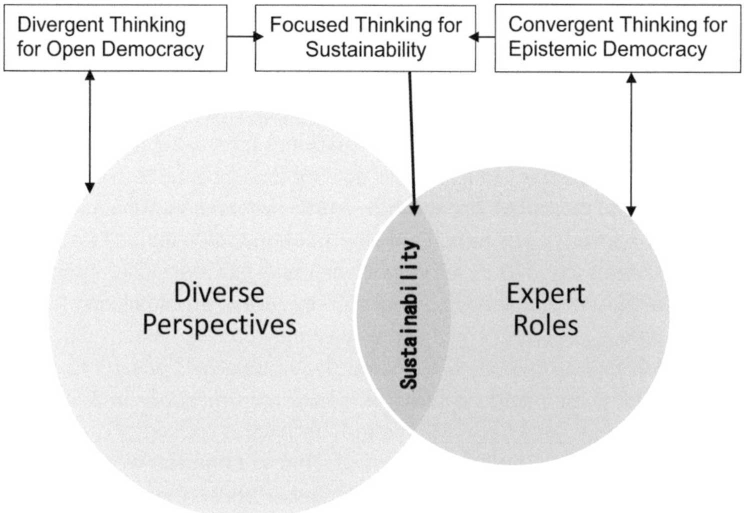

Fig. 5.2 Dynamic epistemic equilibria for sustainability in a democracy

judgements on complex choices as one of the objectives of a political system [24, pp. 183, 232]. Conversely, "open democracy" as conceived by Hélène Landemore is the ideal of a regime in which actual exercise of power is accessible to ordinary citizens via novel forms of democratic representation [49, p. xvii]. However, its five enumerated principles comprised of participation rights, deliberation, the majoritarian principle, democratic representation, and transparency [49, pp. 128–129] do not invoke epistemic conditions as a necessary condition of deliberation by ordinary citizens' in the exercise of their participation rights.[4] To close this gap, open epistemic democracy is advanced to promote collective divergent thinking with epistemic cognitions designed-in to promote collective convergent thinking. The development of any thought system is a continuous dance between the inductive (divergent) and the deductive (convergent) [151, p. 38]. The mechanisms and conditions for epistemic decision-making have been formally modelled by Hong and Page as a "cognitive model" of group decision-making primarily involving the interplay of diversity *and* expertise [152] as interpreted and illustrated in Fig. 5.2.

The tension arising from divergent and convergent thinking moving in opposite directions in the open epistemic democracy framework is resolved at their intersection when sufficient coherence is maintained in divergent thinking, while overspecialisation is mitigated in convergent thinking thereby creating appropriately focused sustainability thinking at which stage the most favourable conditions for transforming thought into sustainable action materialise. It has been shown that neither too much rigidity nor too much fluidity is ideal for cooperation in groups.

[4]The add-on sixth and seventh principles of "dynamic inclusiveness" and "substantive equality" are also subject to the same identified weakness.

Optimal cooperation is attained when there is intermediate fluidity. Among conflicting groups, experiments demonstrate that constructive relations can only be restored if common superordinate goals are in place [28, pp. 106–107, 269]. Indeed, the International Court of Justice ruled in 2025 that states have the duty to cooperate for the protection of the environment and international human rights, as well as to be guided by certain principles (sustainable development, common but differentiated responsibilities and respective capabilities, equity, intergenerational equity, and the precautionary approach). The principal forms of cooperation include financial assistance, capacity building, and technology transfers. The Court also noted that State cooperation is a foundational principle in solving global problems and "it is by co-operating that the States concerned can jointly manage the risks of damage to the environment" [104, pp. 4, 58, 82]. Tellingly, of the 202 democratic and non-democratic national constitutions listed on *constituteproject.org* in 2023, only two contain the word "sustainability" (Somalia with respect to federal governance and South Sudan concerning indigenous games); four contain the word "sustainable" (Fiji with reference to nature itself, South Sudan concerning natural resources, Slovenia with regard to water, and Eswatini concerning "home-grown political order"); and seven others make reference to "sustainable development" (Egypt, El Salvador, Eritrea, Ethiopia, South Africa, Seychelles, and South Sudan). This suggests, among other things, that developed countries are not as focused on sustainability issues as are less developed nations. But more significantly, *191 countries do not currently address environmental sustainability as a constitution-worthy subject.*[5]

In an ideal open epistemic democracy, collective meta-awareness would flow from the interplay of *directed epistemic attention* and *inattention*. Both attention and inattention are precursors to Dan Sperber's notion of *epistemic vigilance* which relies on individual mental mechanisms that are articulated across individuals and populations into social mechanisms. Some of these mechanisms address the source of information, others its content [146]. This is somewhat similar to "pluralistic ignorance", understood to be ignorance about what most other people think. Many social practices persist for this reason, even communism by some accounts, because people were unaware as to how many other people actually despised the system [153, p. 63]. More importantly, miscalculating what other people think also applies to sustainability issues. Research has shown that Americans presume that only a minority of people support a Green New Deal or a Carbon Tax, but in reality the vast majority do [154]. To combat such muddying of the waters, directed epistemic attention is required for *social learning* to occur. When inattention is caused by external factors it may amount to epistemic injustice. We are owed epistemic attention as epistemic agents when we are in a shared common ground; therefore, when epistemic attention deficits arise by rendering our experiences less salient or prominent than they ought to be, they affect our ability to contribute to such common ground, and in a very practical sense, we are then restricted in exercising our epistemic agency. Fluency and range as an epistemic agent depend on the knowledge contributions of others; therefore, structured collectives have an important role to

[5] South Sudan appears under all three positive categories but was counted only once for the total.

play in correcting epistemic attention deficits by ascribing group agency that generates group-level moral duties. In short, the primary harm of epistemic attention deficits is that one is denied something essential to being an epistemic agent or group, and that is a morally relevant concern [155]. Some evolutionary biologists have argued that it is the capacity for moral deliberation *itself*—as opposed to the *contents* of any resultant morality—that has been fitness-enhancing for humans [28, p. 409]. But more simply and powerfully, Iain McGilchrist asserts that attention is a moral act.

Consider further that studies of societal values are largely absent from the mainstream sustainability science agenda. Yet, at its core, sustainability is an ethical concept raising questions concerning the value of nature, responsibilities to future generations, and social justice. Unless those values are articulated and understood, the inevitable political dimensions of sustainability will remain hidden behind scientism and hinder the necessary societal deliberation and convergence on more sustainable pathways. The way values, including resilience, vulnerability, and environmental tipping points are framed need to be elicited and elucidated to support collective decision-making as to what criteria should guide sustainability transitions. An increased focus on the role of values will enable sustainability science to provide information that speaks to common societal values and to advance knowledge on how particular policies or programmes can promote sustainable pathways. When values become a core part of the sustainability research agenda, participatory decision-making approaches would need to follow since decisions would no longer be based solely on expertise and scientific criteria [156]. In societies with low corruption individuals might prosper most by being honest, trusting, and playing by the official rules, whereas in high corruption societies individuals following those norms would be "suckers". The optimal strategy for an individual depends on the strategies pursued by others who chose their strategies on the basis of their expectations about how others will react. Available epistemic information (in the positive case) and asymmetric information (in the negative case) that alter these expectations can change behaviour and may in turn lead to a change in norms that move a group or a whole society from one equilibrium state to another [157]. Dynamic epistemic equilibria as a unified collective decision and action theory addresses this concern by the continuously emergent balance between recognised expertise and informed public participation in decision-making in a particular domain, followed by relevant stakeholder decision-taking at the most appropriate locus.

It is currently problematic that there is no democratically acceptable aggregation method that coherently integrates epistemic inputs across all credible sources for a common global purpose. A founding norm of the UN's Sustainable Development Goals is transversal integration, which is an aspect that has been largely neglected by both national and multilateral institutions. When policymakers take action they typically focus on one goal without due consideration about how it might affect the rest, yet some goals can produce cascading benefits across many of the others. For example, studies show that clean energy can drive progress towards many SDGs without hindering economic development, but this research is mostly ignored. Widespread frustration between what research shows and what policy does is fuelling interest in studying this problem in its own right [158]. To this end, the Global SDG Synthesis Coalition, formed in 2022, comprised of forty UN organisations aims to extract lessons from a multitude of existing studies according to five themes: *partnership*, *people*, *planet*, *prosperity*, and *peace* [159]. Yet this slow-in-coming initiative lacks what should have been the obvious overarching thematic glue of *sustainability.*[6] Along the same lines, a more important opportunity was missed in the drafting of the UN's "Pact for the Future" in 2024 [160] which lacks a transformative reverence for all life on Earth.

Unlike public and private goods, science as a commons is constituted from and by its communities of practice according to the rules that they themselves establish achieving a type of commons which requires joint objectives, deliberations, infrastructures, and rules of use—as well as attitudes based on trust, reciprocity, and cooperation [161, pp. 57–78]. There is an urgent need for more powerful scientific and policy tools for analysing the whole of the integrated Earth system with regularity and reliability to guide political processes so as to prevent altering the state of Earth system beyond levels tolerable for today's societies [137]. Yet as Vaclav Smil has noted, "no government has ever made policies with the biosphere in mind". Although science is far from perfect, it offers our single best hope to find common ground. Professionalism and populism rarely coexist comfortably with each other yet new ways and means must be found to harvest the wisdom from each [162]. We might want to think more open-mindedly and carefully about which particular areas of knowledge deserve to be promoted, which should be left to their own devices, and even which should be actively impeded. Since academics are likely to be biased in favour of thinking that their own field deserves to be promoted, outsiders who are less prejudiced should be brought in to participate in these deliberations. The old model of scientific research, which regards science as a perpetual gift—but only if allowed full autonomy and if shielded from external social control—would perhaps have to be replaced with a different model in which democratic processes and preferences are allowed greater influence over research directions and priorities [157]. One especially hopeful characteristic of science is that it can be leveraged even by individuals and small organisations to produce big impacts. If the audience for science is society at large and not just other scientists, the choice of what data and knowledge one needs has to be informed by the real-world context of the problem to be solved. The questions to be asked are likely to be very different if a solution is sought for concrete problems rather than only to advance understanding [163].

Dynamic Epistemic Equilibria as Fine-Grained Theory

If knowledge is power, knowing what we don't know is wisdom—Adam Grant

Annie Murphy in her book *The Extended Mind* reports that the new science of the group mind shows that we think differently and often better when we think as part of a close-knit group rather than as individuals. This is particularly the case regarding our attention and our motivation, which are altered in meaningful ways

[6] Maybe because the term "sustainability" doesn't begin with the letter "p"?

when we enter them collectively instead of alone. The capacity to be intelligent is not a fixed property of the individual but rather a shifting collective state [13, pp. 11, 221]. The human brain-body system is designed for both learning and behaviour—each of which functions to build an efficient and effective working model which acts as a regulator, encompassing itself, its environment, and the interactions between them. Learning about a complex environment is enhanced by sharing information across individuals with diverse experiences such that social animals have the best chance of accurately modelling their environment and their interactions with it [164, pp. 18–19]. New research has found that when a group of individuals are thinking well together, their patterns of brain activity come to resemble one another's. There is a growing interest in the social dimension of cognition aided by new and adaptable tools[7] that point to the "interactive brain hypothesis". The development of crucial thinking skills such as generating hypothesis, designing experiments, and analysing data has been shown to be highly correlated with the social engagement of one's peers [13, pp. 188, 190, 201]. And from social psychology comes the *self-categorisation theory*, which holds that the self should not be considered as a foundational aspect of cognition, but rather as a product of an overall cognitive system at work. According to this theory, category hierarchies at one level of abstraction resolve social perception conflicts by making salient a more inclusive superordinate ingroup [165].

Arthur Koestler in his book, *The Ghost in the Machine*, devised a conceptual framework with wide applications that he named the *Open Hierarchic Systems Theory* in which a hierarchy of *holons* related to each other make up a coherent structure—a *holarchy.* A holon is something that is simultaneously a whole and a part, a term derived from the Greek: ὅλον, (holon). Holons are the active intermediaries between holism and atomism as configured in a hierarchy. A practical example of what a holon is can be demonstrated by written language. When we read, we do not perceive the form of one letter at a time, but rather the patterns of one or several words at a time. The perception of individual letters is integrated into larger units. Consider the phoneme /e/ which may make up the digram/en/which in turn may be a part of the morpheme/men/[8] [10, pp. 25, 41]. At its earliest, comprehension probably began with the word "men" in this example. The hierarchical configuration between holons at different levels may be said to "make up" the holons or parts of another level, and the levels make up a "holarchy". The otherness between "wholeness" and "partness" is inherent in the concept of a holarchy. Holons are semi-autonomous sub-wholes branching into sub-wholes of a lower order and serve as the functional nodes of a hierarchy. They function as wholes relative to their subordinates, but as dependent parts relative to their superordinates. Holons are governed by fixed sets of rules and flexible strategies that are guided by feedbacks, such as triggers and filters. A social holon displays a self-assertive tendency which is counterbalanced by an integrative tendency in a kind of dynamic equilibrium in

[7] Functional near-infrared spectroscopy (fNIRS) and electroencephalography (EEG).

[8] "men" can also be a stand-alone word, or incorporated into another word or phrase, say, "don't *men*tion it".

coordination with its local environment. The rules of conduct of a social holon cannot be reduced to the rules of conduct of its members [10, pp. 45–58].

Four types of distinct knowledge: *sophia, phronesis, episteme,* and *techne* constitute the core elements in Aristotle's epistemology. The former two make up knowledge aimed at contemplation, while the remainder make up knowledge that aims at action. *Sophia* or theoretical wisdom is the type of knowledge we seek when we want to understand something for its own sake. *Phronesis* makes one excellent at deliberation and involves the ability to grasp the connections between facts, especially when oriented towards action. *Episteme* involves both demonstrative and non-demonstrative scientific knowledge. *Techne* is the skill involved in bringing into being a new object. Typically, we learn cumulatively by building on and combining what we already know as opposed to going from a state of complete ignorance to a state of necessary knowledge. This knowledge is object directed, with "object" encompassing facts, particulars, theories, and ethical norms [166]. The proposed holarchy of dynamic epistemic equilibria illustrated in Fig. 5.3 consists of a hierarchy of three levels with a maximum span of four holons that structurally branches down to the individual level, namely, discrete human beings [31, pp. 320–321]. From both a biological and a cultural perspective, we can theorise further that it is comprised of the accumulated latent knowledge and wisdom of all humanity—an emergent property that only becomes visible when called upon. It is a hierarchical arrangement that dynamically combines the four foundational concepts of learning such that metasophia collectivises techne, episteme, phronesis, and sophia, which are made functional by vertical and horizontal integration of collective tacit and explicit entangled thinking seeded by knowledge. Although the associated bi-directional arrows imply flow, they only hint at the complex dynamic relatedness of all the different processes that occur in different contexts.

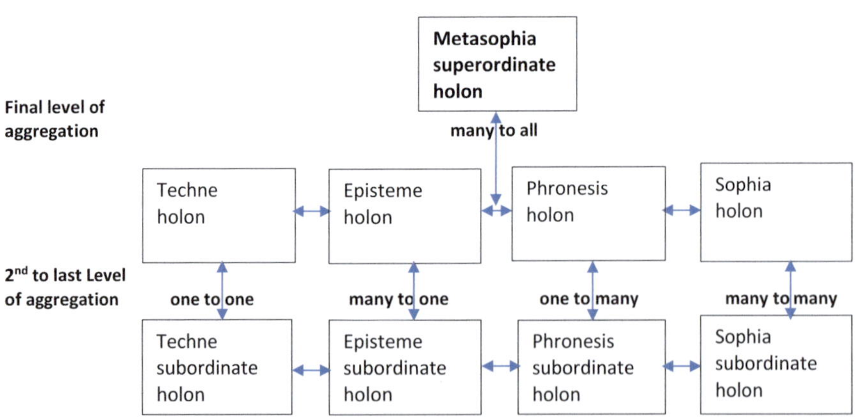

Fig. 5.3 The holarchy of dynamic epistemic equilibria

At this point in theory building, it is plausible to claim that metasophia, or higher-order collective wisdom, has been the last stage of theoretical wisdom all along hidden throughout history behind the four classic formulations of human reasoning leading from techne to episteme and then onto phronesis and sophia. The earlier stage is lifted up into the next stage both in the sense that is subsumed into the succeeding stage, and in the sense that it remains present in, but is transformed by the higher levels of the process [5, p. 204]. Starting with metasophia at the top level of the holarchy, this holon in our schema serves as the structural locus for higher-order wisdom to both emerge and coalesce. It represents the culmination of the collective, conserved, curated, and disseminated knowledge by the *many-to-all*. As the percentage of people taking part in a public conversation rises, so the network becomes more democratic. The only way to have a large-scale political conversation among diverse groups of people is if citizens can gain some understanding of issues that they've never faced themselves. Consequently, in a large polity it becomes a crucial role of the education system and the media to inform people about things they have not experienced firsthand. In the absence of a suitable education system or media platform to perform this role, no meaningful large-scale conversations can take place [76, pp. 135, 142]. The hypothesised possibility of reaching "all" people was strongly suggested by the COVID-19 pandemic of 2020 by the apparent awareness it induced around the globe concerning the wearing of face coverings to thwart the transmission of the disease-causing virus. It would at least be contrary to the precautionary principle to advocate against the many-to-all claim given the prospect for global hazards that failing to achieve environmental sustainability would bring to life on Earth.

At the holon and subordinate holon levels, techne is centred on *one-to-one* aggregation and is epitomised by the reflective practitioner learning from his or her teacher; episteme, as in the term "epistemology" is mainly aggregated in a *many-to-one* mode and is best represented by the many knowledge networks that each one of us benefits from currently; phronesis involves *one-to-many* aggregation as in the individual wisdom of, say, Aristotle when it has been conserved, interpreted, and imparted to many of us, and sophia is best evidenced by the continuity of the philosophy of thought imparted in a *many-to-many* mode from our multiple ancestors and cultures to the many of us. The holarchy's means of integration include channels of feedback that make the whole smarter than any of its parts. These include testability and replicability in science, peer review in publishing and academia, fact-checking and editing in journalism, adversarial proceedings in jurisprudence, and elections and institutional checks and balances in governance[9] [84, pp. 315–316]. Functionally, the holarchy acts as a knowledge network. *Knowledge networks* are self-organised knowledge producing systems that generate new knowledge near

[9] Expanded list based on [84].

spontaneously by various interactions among individuals and groups that are not centrally controlled. These networks consist of the documentation items in libraries, archives, databases, and other repositories consisting of the papers, books, reports, websites, films, and other materials typically produced by scientists, researchers, and authors working in defined fields guided by specific paradigms. These knowledge networks are characterised by concepts, publications, researchers, and information exchange linked by bridging mechanisms, such as published citations, conferences, and online platforms. New knowledge patterns are generated via the non-linear interactions of multiple autonomous agents, while the overall system is functionally a *heterogeneous network* involving three different kinds of nodes: (1) *agents* comprised of individual scientists, groups, and organisations; (2) *containers* consisting of libraries, databases, and documents; and (3) *concepts* made up of abstract knowledge, ontologies, and keyword items [167].

Knowledge networks, however, are also subject to the challenges of complexity ranging from simple ignorance to a lack of understanding of the causal mechanisms involved in large dynamic systems. For instance, the tendency to think in equilibrating mechanisms as opposed to non-linear or chaotic processes in assessing the effects of human actions on ecosystems is counter to sustainability. As long as we assume that a biogeophysical system will revert on its own to some earlier preferred state, we have no compelling reason to adopt the precautionary principle in regulating human actions affecting the system. Under these conditions, the ability of rent seekers to slow or block knowledge legitimacy and progress is significantly enhanced with respect to the resolution of large-scale and highly complex environmental problems [37, pp. 67–68, 80]. When it comes to sustainability and the Earth's resilience, broad coalitions of citizens, businesses, non-profits, and government agencies have the potential to transform how we act as stewards of the biosphere. Science has an important role to play in this context as an honest broker, informing evidence-based action, and in the coproduction of knowledge, all of which are in need of comprehensive renewal in line with anthropologist Margaret Mead's observation that one should "never doubt that a small group of thoughtful, committed citizens can change the world. Indeed, it is the only thing that ever has". The greatest progress of humanity has come from the accumulation of multiple individuals contributing to what eventually becomes a common stock of knowledge. Even the smartest among us influences and is influenced by society, and his or her achievements merely provide variable impetus to move the collective enterprise forward. More specifically, he or she takes what many others have already set out, and adds to it, using methods they have learned from teachers and others that preceded them. No one individual could today master all the complexities of even a single learning domain, let alone space travel if one were to begin from a blank slate [168, p. 137].

Additionally in the proposed schema, a cybernetic process is included such that ongoing learning constitutes a feedback loop. *Re-reflection* using both positive and negative lessons learned can improve outcomes for sustainability similar to what happens in science when both confirmed and disconfirmed hypotheses generally help to advance possible solutions. In open epistemic democracy, this amounts to a process of *re-democratisation* characterised by the agenda-setting powers of sufficiently informed and engaged citizens who select the *path-setting decision criteria* (which is what constitutions are) that lead to intermediate or final *expertise aided decision-taking*. More specifically, the C*ollective/Democratic Decision* Model (see Fig. 10.2) comprised of citizen and expert roles essentially leads to "self-help" democracy with citizens contributing citizen learning, critical thinking, and democratic participation, while experts operationalise the four proposed frames of thought about *emergence*, *relatedness, design,* and *wholeness* (see Chap. 8) in a state space of time-bounded information. When associated strategies and implementation techniques are deployed, in time they can potentially yield higher-order wisdom in the service of sustainability as the foremost collective goal of humanity.

Chapter 6
Democracy as Collective Attention

Freedom is the sure possession of those alone who have the courage to defend it—Pericles

According to Geoff Mulgan, if we regard democracy not only as the communication of citizen preferences, but also as a collective thinking process, then different expectations follow. From this viewpoint, the quality of deliberation matters as much as the quantity of people involved [24, p. 183]. But before the communication of citizen preferences can even be effected, there has to be a *pre-communication* stage otherwise *what* would we be communicating? I claim that pre-communication involves collective attention-making followed by collective attention-taking, the latter honing the former such that a more coherent set of matters comes into view. From a human child pointing as a first mode of communication, to the evolution of pointing to animal trackways by hunter-gatherers, and onwards in leaps and bounds to the crucible of open epistemic democracy, *directed epistemic attention* serves as the overarching democratic attention-making mechanism. Without it, ordinary thought in society would continue to be incoherent going in all sorts of directions, with thoughts conflicting and cancelling each other out; however, were people to think together in a coherent way, it would have tremendous power—coherent movement of thought followed by coherent movement of communication [43, p. 16] that can be aided by critical vigilance. The trajectory of such a democratic path is what Václav Havel has characterised as a "pre-political" stage, he asks:

> What is this independent life of society? The spectrum of its expressions and activities is naturally very wide. It includes everything from *self-education and thinking about the world* [my emphasis], through free creative activity and its communication to others, to the most varied free, civic attitudes, including instances of independent social self-organization. In short, it is an area in which living within the truth becomes articulate and materializes in a visible way [169].

Aristotle in his *Politics* recognised the opportunity for both individual and collective democratic participation when he observed that "some judge a particular part of the issue, while all of them together judge all of the parts taken as a whole". The group achieves its correctness of judgement as a group. Moreover, under the right conditions, "the many" judge certain matters better than any excellent small group or individual [170]. The important lesson to draw from Aristotle's sparse

E. Bellos, *Sustainability Thinking*, https://doi.org/10.1007/978-3-032-11244-6_6

descriptions is to ponder on what the right conditions might entail, and the comparative judgement value of "some" versus the judgement value of "all" or "the many". The *Condorcet Jury Theorem* holds that a majority vote correctly catalogues an issue with a higher probability than any single individual, and as the number of individuals becomes large, the accuracy of the majority vote approaches 100%.[1] The multiple bifurcations of binary choices in the jury theorem assumption of "guilty/not guilty" are one practical limitation to its wider usefulness. More importantly, for the theorem to remain valid, decisions must be arrived at independently with no social influences such as "herding" [150, pp. 15–16]—a caveat which Aristotle did not warn us about—but which is particularly pertinent in today's world of "fake news", conspiracy theories, anti-vaccination sentiment, climate change denialism, and other misinformation. Science is mainly about bringing rigour and transparency into the process of knowledge acquisition and creation. But we have been so inundated by vendors using science and its jargon to sell us products that many of us cannot differentiate science from *scientism* [171, p. 218]. Even more pernicious to healthy collective attention is *Lysenkoism* which has come to mean the manipulation of science to reach pre-determined or ideological conclusions [28, p. 356].

Attention is one of the seven key processes of mind. The other six are perception, learning, cognition, memory, motivation, and action control. Although human attention is sometimes captured by an unexpected sight or sound, it can also be consciously directed and focused, thereby shaping the perceptual stream imposing some organisation on the flow of sensory impressions so that the mind can make sense of what it sees and hears. But mental processing and nerve channel capacities are limited. So why waste time and energy processing and storing unchanging visual inputs or those that could easily be predicted. Information value is the inverse of predictability which means that the more a fact could be predicted on the basis of what is already known, the less learning that fact will contribute to what one knows. In other words, wholly redundant information is no information gain at all [164, pp. 9–10]. What is needed is a wide range of knowledge as suggested by Table 6.1.

Both individual and collective attention can be variously garnered, coalesced, and redirected in aid of sustainability. The stock of governance tools and ways to adapt and combine them is far greater than is often recognised. Informational systems that meet high scientific standards and concurrently serve the ongoing needs of users and decision-makers are particularly useful. Well-structured dialogue involving scientists, resource users, and affected publics who are informed about the workings of human-environment systems results in analytic deliberation that focuses relevant information that can produce consensus on governance rules [56]. And if attention were not as important as suggested here, it would be hard to explain from a bottom-up perspective the herding, flocking, and schooling behaviours in the animal kingdom which evolved precisely to thwart the focused attention of

[1] Scott Page (2018) applies the Condorcet Jury Theorem to argue for the application of many models to a problem, which is a valid proposition because inanimate models can be made to be independent, but people are subject to social influences and therefore cannot act truly independently.

Table 6.1 Types of knowledge in epistemic making

Tacit knowledge	Common knowledge
Know *how*	Know *what*
Techne	*Episteme*
Practice	Scholarship
Non-demonstrative knowledge	Demonstrative knowledge
Experiential knowledge	Epistemic base
Heuristic	Propositional knowledge
Observational	Rational
Figurative	Literal
Ideation	Customary

Adapted from [33, p. 467]

predators. But more to the point, Havel's "independent life of society" can be improved by four attention building mechanisms: *a learning culture; Bayesian priors; Bayesian posteriors;* and *directed epistemic attention.*

First: A Learning Culture

> *We have inherited from our forefathers the keen longing for unified, all-embracing knowledge*—Erwin Schrodinger

Learning is innate in humans. Consider that healthy humans have immune systems that can learn in a process similar to that of evolution but on a time scale of days instead of decades or more. Human infants can acquire a native language, an inborn ability supplemented by an internal grammar at the end of the learning process that permits a child to tell whether a previously unheard sentence is grammatically correct or not [17, pp. 19,52–53]. And as we saw earlier, all humans have innate curiosity [10, p. 153]—curiosity is both the unconscious and conscious substrate for our continued drive to learn and branches out from that part of our common gene pool to reach every one of us. In his book *Blueprint*, Nicholas Christakis argues that genes can explain why human culture exists in the first place. Evolution provides the substrate for human culture by equipping us with the ability to cooperate, make friends, and learn socially. Human culture is a property of groups, not individuals whereby culture affects genes, and genes affect culture. This set of interactions is known as the theory of gene-culture coevolution. Cooperation is an essential predicate for social learning, one of our species' most effective inventions. No human has to learn everything on his or her own—we can all rely on others to teach us—a highly efficient practice present in all cultures. But the crucial attribute of our species is our ability to fashion diverse cultures from our pre-wired inclination to do so. This capacity for culture, rather than its by-products, reflects our evolved proclivities for social interaction, cooperation, and learning [28, pp. 306–310, 365–370].

There are many ways to frame a discussion on the role of learning in human affairs. Notably, from physics to medicine, reliance on experiments has taken many forms and played a critical role in shaping human thought and learning. Here we are mostly concerned with how experiments as both mindset and technique can be leveraged by governments, experts, and citizens to promote social learning. Problems materialise when governments have to confront the uncertainty of complex challenges by adapting or creating entirely new service systems in the context of path dependence. Experimentation in science and government is often equated with doing Randomised Control Trials, which are an important part of qualifying and validating a hypothesis, but they are less useful if the understanding of the problem is unclear and the solution space needs to be re-imagined. Experimentation helps fill these knowledge gaps by systematically testing assumptions without allocating too many resources, or much time. Ideas never emerge fully formed, but rather have to be adapted to a dynamic system. Reframing ideas as hypotheses draws attention to the need for ongoing testing, refinement, and further development. In organisations this means reframing pre-determined objectives from an almost exclusive focus on upward accountability to more deliberate downstream learning [172]. Generally speaking, moving towards an experimental culture in organisations can have multiple and emergent benefits.

Methodologically, experimentation can be seen as a way to traverse uncertainty towards more calculable risks by making one decision and then building on it. This is more important in highly complex environments where it is essential to keep the mandate for an experiment independent, that is, not tied directly to policies and strategies because the results may require that a particular course of action be abandoned. This also helps to keep the original scope of the experiment intact. Many social interventions cannot be assessed feasibly through experimentation due to ethical or other reasons, and in some cases, replicability, which is a hallmark of traditional science, is not practical [96, p. 89]. A summary of conditions favourable for the emergence of experimentation and social learning in public affairs is presented in Table 6.2.

Experimentation is often a key contribution of public sector innovation teams and government policy labs. Organisations like *Lab para la Ciudad* in Mexico City, *Alberta CoLab* in Canada, *Behavioural Insights and Design Unit* in Singapore, *MindLab* in Denmark, and *Policy Lab* in the UK are specifically set up to promote, develop, and embed experimental approaches for public policy and to accelerate user-centred learning at different levels of government. Experimentation can occur on a continuum between exploration and validation involving the generation of hypotheses, establishing hypotheses, and validating them [172]. *Experimental Finland* is similarly guided by a three-phase approach consisting of qualitative experiments, verificatory experiments, and evaluation of experiments [173]. Between exploring options and validating a hypothesis along the continuum of experimentation there is also a "trial-and-error" approach that builds on both the imaginative and analytical mindset. A bricolage of experimental methodologies from different disciplines such as data analytics and design can serve to underscore that successful experimentation needs a dynamic and iterative process so as to

Table 6.2 Some existing and proposed elements of a learning culture[a]

Traditional learning culture	Traditional policy and programme planning	Emergent learning culture	Experimental policy and programme planning
Randomised control trials are relied on in medicine and the natural and social sciences	Outcomes are predictable and result from careful (process) execution	**Outcomes are assessed using multiple criteria**	Outcomes are not predictable and result from trial-and-error
Disciplinary boundaries are defended/maintained	Existing knowledge/methods are used to achieve desired outcomes	**Inter- and transdisciplinary methods are encouraged**	Testing and seeing what works, or tapping into other sources of knowledge (citizens/partners)
Pre-determined objectives are set by central planners	Planning is done for creating a specific future	**Joint expert/citizen scenario planning**	Planning is done for multiple possible futures
Before-after quasi-experiments	Decisions are based on historical facts and past successes	**Knowledge generation creates cycles of innovation**	Decisions are open-ended and based on indications from experiments
Overreliance on experts	Citizens are seen as a problem to be solved or the cause of the problem	**Citizens are a part of the solution**	Citizens are a source of insight and potential for transformative action
Overreliance on static quantitative and qualitative methods	Knowledge is used to make generalisable assumptions on societal problems at scale, at a point-in-time	**Bricolage of quantitative and qualitative experiments include human behavioural components**	Knowledge is used to capture human experience/social complexity and to test assumptions in real-time over time
Feedback from implementation is not incorporated in planning	Policy documents describe or prescribe a predefined course of action	**Innovation is incentivised by competition, prizes, and other means**	Innovation tools enable creative thinking, collaboration, and learning
Overreliance on existing stock of knowledge	Determination as to what course of action to take is made from behind a desk	**Participatory research methods are incorporated in experiment's design**	Consultation with users and stakeholders to determine what course of action to take
Transparency is weakened by jargon and time-boundedness	Uncertainty and risks are camouflaged via detailed plans	**Long-term and intergenerational consequences are considered**	Uncertainty and risks are acknowledged and mitigated
Failure is not viewed as opportunity	Failure is seen as something to be avoided	**Citizen critical thinking and learning are encouraged**	Failure is accepted as part of iterative learning and improvement

[a]Items in **bold** represent proposed reconfigured elements

Author based and inspired by Nesta (https://www.nesta.org.uk/blog/towards-an-experimental-culture-in-government-reflections-on-and-from-practice/)

influence direction and for creating a basis for redesign and legitimising decision-making [172].

Next: Bayesian Priors

We are who we are because of what we learn and what we remember—Eric Kandel

From physics we learned earlier of the *principle of least action*. In human behaviour we speak of the *principle of least effort*. Effortful attention is needed to combat the principle of least effort. All life, including the substrate of our thoughts and memories, is made up of the same building blocks. Selective attention is widely accepted as a powerful factor in perception, memory, and action—in the wholeness of conscious experience—it is critical to the unitary nature of consciousness. Attention acts as a filter. It is in large part due to selective attention that mental representations do not replicate every detail of the external world, we focus on specific sensory information and more or less exclude the rest. A modulatory pathway in attention-related events is mediated by dopamine (a hormone and neurotransmitter). The cells that manufacture dopamine are clustered in the midbrain, and their axons send signals to the hippocampus and prefrontal cortex areas of the brain adjusting the firings of these neurons for both voluntary behaviours and attentional processes. According to William James, in his book *The Principles of Psychology*, there are at least two types of attention—involuntary and voluntary. Involuntary attention is supported by automatic neural processes and is particularly relevant in implicit memory, for example, when something has proven to be salient or surprising. Involuntary attention is activated by "big things, bright things, moving things, or blood". Conversely, voluntary attention, such as reading what one is writing, is a specific feature of explicit memory that arises out of an internal need to process stimuli that are not automatically salient so as to withdraw from some things in order to deal more effectively with others. The controlling factor in whether memory is implicit or explicit is the way in which the attentional signal for salience is recruited—from the bottom-up or from the top-down, that is, reflexively or voluntarily [174, pp. 236, 311–314]. Also of some help are the various traditions of mindfulness that teach how to attend to attention itself—to be conscious of the ways in which unwanted thoughts can crowd out other thoughts [24, p. 37]. For routine activities we can add whether attention is from a wide-angle or a narrow focus as suggested earlier by the mixed-scanning model.

In narrow focus, when every tiny detail of an experience is remembered in a rare condition known as *hyperthymesia*, memory becomes a terrible burden. The fact that hyperthymia exists at all indicates that the human brain is capable of storing a massive amount of information not unlike modern computers. It would seem that this storage capacity would be an advantageous capability, but not in light of evolution's preference for our ability to act effectively. As brains got more complex, they got better at responding to deeper, more abstract inputs from the environment which makes them ever more adaptive to new situations. Storing too many details is often unnecessary to act effectively, it can be counterproductive for choosing the most useful actions that would enable survival in the long run. Remembering everything would get in the way of focusing on the deeper principles that allow us to compare how a new situation resembles similar past situations, and in consequence, what kind of actions would be the most effective [147, pp. 38–39, 47–48]. Too much focus and too little peripheral vision constrain us in picking-out the novel pattern, the unforeseen threat. Interestingly, professional shooters use lenses that reduce blue light with a filter to improve clarity—by seeing less, they see more [24, pp. 37,143].

There is, of course, also the possibility of not paying attention at all. Our neglect of *priors* is especially important in two cases: *Bayesian priors* and *base rate neglect*. In the case of Bayesian priors, learning links directly to the guidance of behaviour by making expectation the centre of learning. Bayesian learning is about the efficient extraction from ongoing experience of probabilistic information used to update model-based expectations. A record of past events, however accurate, is silent about what will happen next. Mental Bayesian processes are, in effect, running summaries of the impact of past information thereby affording instantaneous access to the final outcome of what one has learnt without requiring an extensive search of memory to determine the relative weight of the evidence. For humans who ordinarily deliberate, using intuitively derived strengths of belief or degrees of uncertainty, such processes suggest a mechanism by which such intuitions could be sensitive to what we have learnt implicitly, as well as explicitly, and how reliable these has been.

Our second class of neglecting priors is *base rate neglect*. Neglecting base rates is the tendency to regard a particular case as being representative of its whole class without first referring to a baseline [84, 85, p. 249]. This means that a prior probability, that is, an event that has already materialised is going to be a more accurate predictor than a base rate, especially if the rate was derived from a non-random sample. Add the question of *which* base rate should be used and the choice of an appropriate *reference class* becomes critical [84, 85, p. 249]. For example, for my estate planning, should my attorney look at Quality Adjusted Years of my age group, or a combination of other socio-economic factors? She would be forced to choose a relevant reference class because no single event or prior probability assessment of mine would be objective enough on its own to make estate planning decisions on my behalf.

Then: Bayesian Posteriors

We must urgently learn to live differently—UNESCO

In the case of Bayesian posteriors, if expectations are consistently modified in the face of experience in a Bayesian manner, then over time, the influence of initial expectations will tend to diminish as new experiences fine-tune expectations to actual occurrences by the reduction of prediction error. Importantly, they will tend to converge on the actual "natural statistics" of their environment [175, pp. 63–64]. More colloquially: now that you have seen the evidence, how much should you believe in your idea? This involves re-examining your credence in the idea/hypothesis/issue after taking into account the visible evidence for the hypothesis, which when multiplied by the likelihood of the occurrence of that data, scaled by how prevalent that data is more generally, your beliefs can be systematically updated [84, p. 153]. If Bayesian inference is the normative standard, updating our beliefs accordingly would be optimal for minimising our long-term inaccuracy. But sometimes we just want to be mostly right quickly, sacrificing some amount of accuracy for efficiency, even if it means getting something wrong in some cases [81].

What might be the group correlates to individual attention? By way of biology, research on colour suggests that exposure to 480 nanometres before 11:00 a.m. stimulates attention. In this range of the light spectrum, what we see is the colour *blue* [24, pp. 37,143]. Is there a practical application to this phenomenon? The timing of public events comes to mind. Common knowledge can be created by *focal points*, public happenings in which people see other people seeing. Collective coordination is dependent on common knowledge by being that something that everyone knows, that everyone knows, ad infinitum [84, pp. 124–125]. But common knowledge need not be ascertained only by reasoning through other people's beliefs about other people's beliefs and generalising inductively to an infinity of levels. Instead, we can infer it by using a variety of perceptual and conceptual cues, including announced messages, public rituals, eye contact, inadvertent speech, focal points [176], or some combination of these. A public outrage can arise from focal points and serve to get a group to act together when an atrocity peaks while the right moment to deal with it seems to never come. The real or perceived affront can trigger spontaneous indignation even in a dispersed constituency and forge them into a resolute collective regardless of the amount of harm that was actually inflicted [84, pp. 124–125]. People are highly sensitive to the distinction between common knowledge and mere private knowledge and strategise using this knowledge about knowledge in diverse social situations, such as bystander (non)intervention and moral condemnation. The game theory of coordination predicts that people's sensitivity to common knowledge is ubiquitous in social life and underscores the distinctly human obsession with publicity, privacy, salience, secrecy, outrage, shame, hypocrisy, discretion, and taboo [176].

Now: Directed Epistemic Attention

Art has something to do with an arrest of attention in the midst of distraction—Saul Bellow

Attention has an important role to play in influencing people's beliefs and directing what they take to be epistemically, politically, or socially relevant. For particular views to have any influence on what others believe they first need to be paid attention to. Not doing so, restricts epistemic agents from contributing to the shared common ground, the ability to set the agenda, and the opportunity to influence people's beliefs. Individuals may disagree as to what items should count as common ground, which itself may contain falsehoods, political myths, implicit prejudices, and even intentional propaganda. Nonetheless, if they are generally believed to be "known", or when such items can be used without widespread disagreement, they become common knowledge. But when we pay undue attention to some perspectives and less to others, the common ground naturally shifts. Major contributions to the shift come from powerful media agents who direct attention and engage in agenda-setting, effectively determining what the common ground of discourse is. By deciding what stories to cover, how to cover them, and what prominence to give them, mass media have the ability to influence what issues are seen as important much more so than ordinary citizens can. Where the media direct our attention can determine what counts as consequential or relevant. The importance of agenda-setting is clear: those who set the agenda determine both what counts as knowledge and the social and political spheres worthy of engagement. When we fail to provide due epistemic attention to certain individuals or groups, we not only miss out on specific contributions to the shared set of common knowledge, we also deny agential inclusion and the opportunity to influence those with different opinions. Journalists, educators, and others who work to direct people's attention towards certain viewpoints can help guide attention for less powerful voices. Cultural producers similarly have an important role to play. Artists, poets, novelists, musicians, and filmmakers can also help to balance our patterns of attention. And political organisations and social movements are among the most influential players in changing the dominant patterns of collective attention[2] [155]. At a measurable high-point of potential influence are nation-states periodically ranked via the *Soft Power Index*. Yet nation-states may no longer reflect the only or proper level at which citizens have allegiances, as well as prove inadequate sites of decision-making in the face of transnational or global collective action problems [49, p. 213].

An obvious source of modern actionable knowledge is the very same public that collective decisions will eventually affect. According to the Organisation for Economic Co-operation and Development (a multilateral organisation of advanced economies) stakeholder maps have been redrawn. Citizens are now located at, or near the centre where focus must be placed. Public services that are unable to adapt to citizen participation will need to be fundamentally reworked. Processes that are not meaningful to individual citizens may struggle to build coalitions of support when citizens expect ever more personalised services despite increasingly diverse country populations. Standardised, large-scale public service delivery by means of command-and-control administrative systems no longer functions well, forcing governments to rethink service delivery boundaries and to design solutions that take into account a broader set of actors and their relationships [96, p. 17]. Here we are interested in the attention paid to people as epistemic agents, including the attention

[2] Summary of the referenced article from a positive perspective.

paid to citizens in their role as likely sources of knowledge that can be reciprocally acknowledged, developed, and celebrated. Being an epistemic agent means being able to *know*—with some degree of accuracy, capability, and credibility—thereby raising one's ethical status to maintain continued participation in deliberations [155]. Thus, both moral and intellectual capital in the form of reputations are *first* for participation in an open epistemic democracy.

Laws and other artefacts of society contribute in different ways to collective attention-making. Signalling is one such mechanism but one which is relatively non-directed when compared to, say, election campaigns. One line of research has shown that voters that have been prompted and made anxious display more learning and attention to campaigns, while calm voters pay less attention. Voter anxiety is raised when a stimulus is found to be incongruent with expectations causing a shift in attention. This interruption results in active processing, where both attention to the new information and the time taken to process it increase [177]. This implies that if no unexpected information is supplied in political advertisements, little to no new attention will be garnered. According to one study, the best estimate of the effects of campaign advertising and contact on candidate choice in American general elections is zero. Persuasive effects only emerge in two rare circumstances: when candidates take unusually unpopular positions, and when campaigns invest unusually heavily in identifying persuadable voters [178]. By some other accounts, however, candidates can receive a vote for approximately every $10 they spend [127] while "deep canvassing" has been credited with adding three votes for every one hundred completed phone calls to Joe Biden's 2020 presidential campaign [179]. But these election-related findings miss the larger point as to the *whole value* of candidate campaigns. They are attention-making regardless of their success in changing voters' selection of candidates. Any new information bearing ideas create possibilities for different courses of action which would not have otherwise been possible in the absence of those ideas. Stated differently: negative ignorance becomes knowledge [10, p. 199] and properly developed can serve as a positive force to accompany directed epistemic attention. For example, Andrew Yang, a 2020 U.S. presidential candidate, proposed that data generated by each individual citizen should be considered a property right and be owned by them. Interestingly, his policy proposal would enable individuals to share in the economic value stemming from their data which could be even greater if consumer opt-in options were to be invoked. Data generated during the normal course of our daily activities are very valuable and are increasingly collected by private corporations often without our knowledge, consent, or adequate protections. This policy would also enhance the capability of individuals to keep their privacy intact while using modern devices like cell phones and the internet [180]. Bringing data as a public issue and defining them as a property right simultaneously creates three fundamental paths of causality into view. These three origins head in directions that have been termed "Chain", "Fork", and "Collider" by Judea Pearl as shown in Fig. 6.1 and as applied to Yang's proposal.

A candidate policymaker interested in the *policy implications* of property rights would likely channel her attention to the Fork option because property rights *directly*

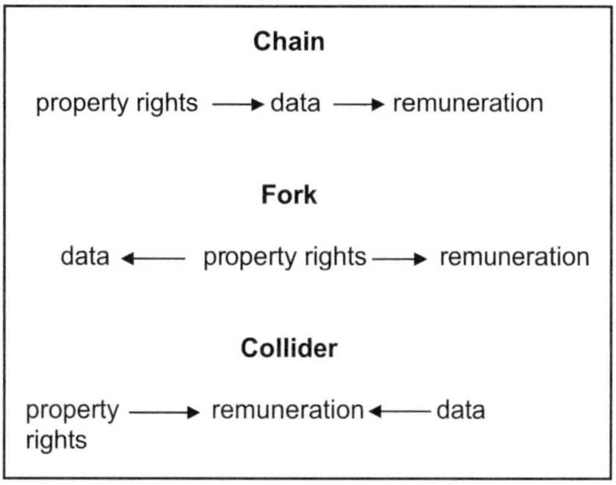

Fig. 6.1 Indicative attention-taking paths for personal data policies

affect data *and* remuneration as the first take of many possible subsequent decision-making nodes and branches. However, very different directions would present themselves to a candidate policymaker more interested in reaching remuneration as a *policy target* who, in consequence, would be more likely to focus on the Collider option because both data and property rights can *independently of each other* result in remuneration. We can enrich our analysis by recognising that in this formulation policy implications represent a divergent outlook, while policy targets represent a convergent outlook, both of which can be applied to collective decision-making as discussed earlier. Yet the average voter may have a fundamentally different perspective from policymakers and is more likely to be interested in a truncated Chain of causation, that is, from her personal data to her remuneration without according any attention to property rights as the root cause of her good fortune. Thus, *different perspectives lead to different attention-taking outcomes subject to existing or adopted roles* (policymakers and citizens in this instance). Our intrepid voter will also have a tendency to conform to what others are doing, to follow local norms, to privilege high status or older people who could be potential teachers, and to pay attention to the people whom others are paying attention to [28, p. 368].

It follows that *variable attention budgets* at a minimum subsisting of personal interests, motivation, memory, and context variables are also needed for recursive joint individual attention to influence public policies. Attention budgets are counterweights to Antony Down's "rational ignorance" which holds that if one has no incentive to cast a costly vote (time, energy, and so on), one also has no incentive to gather costly political information. But what might be seemingly rational at the individual level becomes collectively irrational if all voters are ignorant by default because effective democratic accountability would then be absent and political collapse would eventually ensue [181, p. 297]. This implies that we should economise on citizens' time and attention out of consideration for their own private pursuits, as well as the upward and lateral expansion pressures suggested by open epistemic democracy. To the extent that the new model ultimately involves decentralising decision-making at lower levels, expanding democratic decision-making at supranational levels, and extending democratic decision-making into spheres currently not considered political, such as multinational corporations, it is essential to consider citizens' time and attention as scarce resources [49, p. 206]. These scarce resources can be either absolute or relative. In particular, *epistemic inattention* against some members of a group allows others to have greater, potentially unfair, epistemic attention in a given domain. Those who remain silent in response to the inattention in such situations may enter into a negative spiral of non-participation (we do not pay attention to those we have become accustomed to not hearing from) who are at the same time deprived of the opportunity to directly influence the take-up of their knowledge or positions further distancing them from the shared common ground. This harm can be exacerbated even more because those who would set the agenda in one's disfavour have a greater relative impact than they ought to [155].

A recent study found that when there is a perceived breakdown in the social fabric, heightened moral polarisation predicts increased support for authoritarian-style "strong" leaders, while a perceived breakdown in leadership itself leads to increased support for democratic-style "strong" leaders [182]. For example, a 2021 study found that priming the attention of American individuals towards news headline accuracy increases the quality of the news that people subsequently share on social media. Irrespective of political alignment and despite their strongly shared preference for accurate sharing, misinformation was often shared because their attention was focused on factors other than accuracy. Although cross-cultural generalisability has not been established, social media platforms could easily implement attention-based interventions to counter misinformation online by, among other things, periodically asking users to rate the accuracy of randomly selected headlines thereby subtly reminding them of the importance of accuracy without engendering reactance. Such interventions could over time improve the quality of news circulating online without relying on a centralised institution to certify truth and censor falsehood [183]. And it is important to keep in mind that truth is something that brings our attention to certain aspects of reality while inevitably ignoring others [76, p. 10].

For epistemic attention deficits to be substantively addressed, a societal effort needs to be made to change the patterns of epistemic attention. These patterns of attention form part of our social practices and are substantially informed and in turn, inform the dominant ideology of society. A stable political regime involves patterns of attention that help to stabilise order and form the *attention regime* of a society, which more likely than not serves the interests of the more powerful [155]. In response, attention-making and attention-taking can by design take many different educative forms. Examples range from the mandatory *gram sabas* of Indian villages colourfully described as "verbal democracy", to the National Public Policy Conferences of Brazil aggregated across spheres of government, and the *vusela* national consultations in Eswatini (Swaziland) which converted a traditional monarchy into a constitutional monarchy. In 2012, Icelanders rewrote their constitution involving a randomly selected forum of 950 citizens, an elected assembly of twenty-five citizens, and crowdsourcing techniques to attain even broader public participation. In 2019, the German-speaking community of Belgium established a randomly chosen Citizens' Council to help the elected Parliament formulate legislation. And given until 2020, France's Citizen Convention on Climate Change was mandated to prepare draft laws on all issues relating to ways to combat climate change [49, pp. xiv–xvii]. Ongoing efforts to concentrate democratic attention include Estonia's *Rahvaalgatus.ee,* which is a new web platform that enables citizens to co-create, sign digitally, propose to the Parliament, and then follow the processing of collective proposals.

Beyond elections, mechanisms such as rights of referral, referenda, model laws, citizen initiatives, and providing government solicited inputs open pathways to more just societies.[3] This means that policy proposals need not come exclusively from politicians to citizens, but should also be directed from citizens to politicians. For instance, citizen initiatives serve in part a remedial function, to compensate for the all-too-common blind spots of legislatures, and are also useful methods for citizens to put on the agenda propositions that, through commission or omission, would not have been considered otherwise. In particular, these devices can force classically elected representatives to improve arrangements for which they have no incentives to change, or may have even put in place for the own personal benefit. Initiatives are all about allowing citizens to shift the balance of power between the represented and their representatives thus preventing entrenchment and atrophy that ultimately results in the closure of power. Such tools with supporting procedures would prove particularly useful in removing political obstacles in the transition to more open democracy. Conversely, a lesson from Ancient Athenian democracy is that political rights come with responsibilities. The *graphe paranomon* procedure was a kind of built-in democratic judicial review that ensured that citizens proposing new laws did their homework to confirm that their proposals were in keeping with the fundamental laws of Athens (now we would say "constitutional") and could be kept accountable to this end [49, pp. 75–76, 101, 206]. Notwithstanding the guilty verdicts in 2023 of some "January 6" insurrection leaders in the United States, sanctioning citizens would be less acceptable these days and this need is generally obviated for well-meaning citizens given that sensibilities have been increasing while the costs of acquiring the necessary information for democratic participation have been decreasing. As a means of two-way communication, the Internet provides voters the ability to speak to their governments without leaving their homes. Unlike television, very low technological and financial barriers exist to establishing a presence on the Internet where it is already available, or in relatively short time lags for places awaiting its arrival [127]. Similarly, most people expect digital surveillance to increase, however, gaining public support for its use would require, among other things, ethical and transparent protocols on the collection, ownership, and use of personal data. Learning and understanding from historically and culturally diverse experiences of value-laden issues such as digital surveillance will be a vital input to emerging systems designed to regulate and oversee their use [107].

By way of synthesis, Fig. 6.2 visually organises the underlying conditions necessary for democratic recursive attention-taking to materialise.

[3] See Appendix I for initiatives about initiatives.

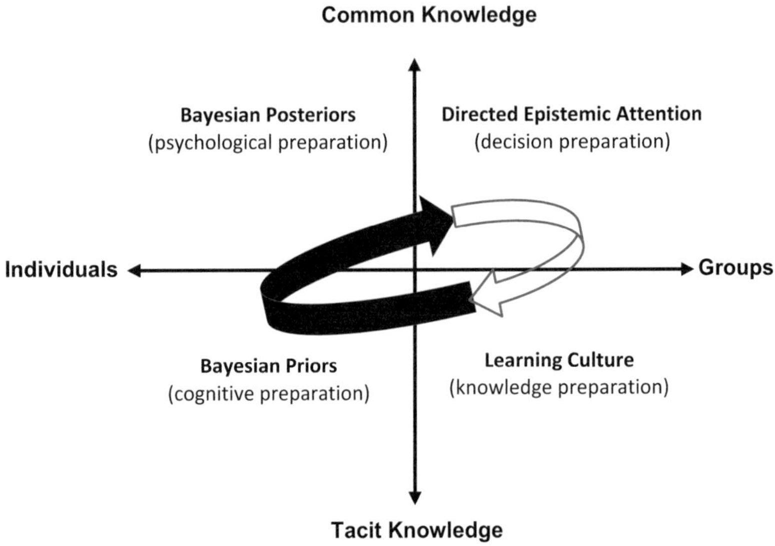

Fig. 6.2 Conditions for democratic attention-taking

Starting from the bottom right-hand corner in Fig. 6.2, there is a precursor phase that conjoins with the other three leading to aggregate attention-taking. Our worldview is formed here, is deeply felt, unquestioned, and unstated. These unconscious assumptions about how the world works set the boundary conditions within which institutions and technologies are designed to function. Institutions such as education and the media play a critical role in shaping our goals and worldview [127]. Cultures and selves define and build upon each other in a continuous cycle of mutual constitution. The self is the "me" at the centre of experience—an ongoing sense of awareness and agency that guides actions and takes shape as both brain and body adapt to various environments. People are socio-culturally shaped shapers of their environments—they make each other up and are in consequence best analysed together. Selves incorporate the patterning of their various environments conferring culture-specific form and function to the psychological processes they organise, such as attention and motivation. In turn, as selves interact with their socio-cultural contexts, they reinforce and sometimes change the ideas, practices, and institutions of these environments [184]. Due to the ability of groups to socially interconnect and learn from one another, cultural evolution has generated collectively held moral sentiments and meanings, which serve as foundations from which formal and informal institutions have emerged [50]. At this starting corner, common knowledge is diffuse, at an early developmental phase.

Moving counter clockwise in Fig. 6.2, Hebbian learning [185] at the neuron level builds up intuition that contributes to the formation of *Bayesian priors,* which are the conditions explained by individuals' subjective assessments of the likelihood of something happening or not based on their prior experience and knowledge. Priors also draw from visual perception and interpretation at the unconscious level. This synthesised state is tacit and helps the individual to formulate his or her position on public issues (among other things) in our schema. Research has shown that emotions play an important part in most decision-making realms. While classical political theory drew distinctions between reason and emotion (affect), interdisciplinary connections from neuroscience to political science have established that emotions must be an integral part of political decision-making processes. Newly encountered information carries with it an affective value. Given an existing assessment, these affective elements interact so that the baseline tally directly influences how the new information is evaluated *before* it is used to update the baseline. Growing evidence suggests that voters may operate with processing biases as motivated reasoners attempting to hold onto their existing positive assessments to explain away new incongruent information. Although voters are psychologically motivated to do so, they reach an *affective tipping point* (despite negative information being weighted more than positive information) implying that voters might be open to disconfirming information after all [177].

At the top left corner of Fig. 6.2, *Bayesian posteriors* take into account actual data surfaced from the general availability of supporting data, which by definition are also available for others to see and assimilate as needed. It is instructive to turn to history to see how people's *shared* Bayesian priors may also become one's Bayesian posteriors in modified or unmodified form. Human ingenuity did not have to wait for the advent of Bayesian statistics in the eighteenth century to deploy epistemic democracy. A democracy may be regarded as "epistemic" to the degree it employs collective wisdom to make good policy. The city state of Athens, from the late sixth through the late fourth century B.C., has been intensively studied as a case study of a participatory epistemic democracy. Ancient Athens relied at times on initial attention-making. One such source was the oracle of Apollo at Delphi, which was consulted twice in rapid succession by the Athenian embassy that was sent to ask what to do about the invading Persians in the fifth century B.C. A local woman named Pythia was the medium by which cryptic oracles that amounted to tests of self-knowledge and insight were delivered. When the Athenian Assembly was convened to debate the meaning of the oracles, Themistocles took the opportunity to interpret the oracle's reference to a "wooden wall" as being a floating wall of ships. The majority of the Athenian Assembly was swayed by his hard-won interpretation and in consequence committed all of its manpower to its navy, leading to the ensuing victory over the much larger Persian fleet at Salamis [7, pp. 29–42]. We can see from this early example the outlines of attention-making, attention-taking, and finally to policy selection and implementation. Other enabling conditions for epistemic democracy included the reforms of Cleisthenes, in particular the Council of 500 in the sixth century B.C. This body of citizens chosen by lot became a learning organisation over time with enough social capital gains to overcome its collective action problems and to work towards a common goal. Witnessing the increased gains that increased communication brought within the strong-tie network within the Council served to loosen the proprietary information of experts upon the realisation that their tacit knowledge was valuable when brought to the surface and shared within a diverse group of people who possessed valuable tacit knowledge of their own. This diversity was institutionalised mainly by term limits of one year. And because each Councillor's also had a network of contacts outside the Council, each Councillor served as a weak-tie bridge to a local subset of the larger population. In consequence, as a body, the Council could gain access at fairly low cost to a good deal of the common knowledge available potentially up to knowing that "Athens knows what the Athenians know". So, Athens' success relative to its competitors can be attributed in part to innovative institutional design, the individual sophistication of its participating citizens, and the diversity of their perspectives [186].

The top right-hand corner of Fig. 6.2 represents the combined attention-taking of both individuals and groups, Properly fashioned, any number of modern democratic processes and outcomes can be "co-managed" by informal means at one end of the range and formal accountability reports at the other when they are used as a part of the policymaking cycle. In the open epistemic democracy conception, these can include citizen initiatives, deliberative polls, "citizen juries", and Delphi-like citizen participation methods, among others. Taken together with all their permutations, Bayesian priors and Bayesian posteriors may constitute a *Bayesian network* made up of various types of knowledge. In other words, when democracy itself becomes *metabolised collective thought*. At the individual level, there is a critical role and duty to act as an epistemic agent in an open epistemic democracy. One way to promote such conditions is to design overarching convergent and divergent mechanisms to guide directed epistemic attention, which in our schema focuses recursive individual and collective attention of both politicians and citizens on the goal of ensuring Earth's sustainability.

Chapter 7
Citizen Roles

A Nation's best defense is an educated citizenry—Thomas Jefferson

In human evolutionary terms, collaborating towards a joint goal has created a new kind of social engagement, a joint intentionality in which "we" (whether hunting antelopes or raiding a neighbouring village), with "my" own interdependent role together with yours produce a dual-level structure of simultaneous sharedness and individuality—*a joint goal but with individual roles*. This calls for relational thinking by functional categories—a uniquely human form of second-personal joint engagement requiring species unique cognitive skills and motivational propensities. Humans have become exceptional at this by creating role-based categories such as archer, runner, wife, teacher, beneficiary, expert, and so on as part of some larger activity [27, pp. 42–43]. Roles for humans are essentially *cognitive divisions of labour* [147] underpinned by our biology. The human ability for sequential planning, the use of extended time horizons, and prospective imaginal simulations enable us to provisionally project ourselves into temporarily distant scenarios. Indeed, there appears to be a set of interconnected regions of the human brain, called the *default network* specialised for this purpose. Areas of the default network have undergone extensive evolution in the transition from mammalian ancestors to modern humans. Our expanded abilities for prospective simulation play a key role in our capacities for sequential planning and our use of distant time horizons when compared to our animal friends [187, p. 195]. More succinctly: human beings *metabolise* the past to simulate the future thereby creating new thought that permeates both conscious and subconscious processes [188, p. 352].

From a child learning a human language to a mid-career professional learning a new coding language, the role of *learner* is pervasive in modern society. Neri Oxman has captured this reality by way of a graphic (Fig. 7.1) modelled on the Krebs cycle of energy conversion in biology that we might call "knowledge metabolism". It is an attempt to illustrate the "antidisciplinary hypothesis", namely, that knowledge can no longer be ascribed to, or produced exclusively within, academic disciplinary boundaries because it is now entirely entangled. Using four proposed domains of knowledge exploration—science, engineering, design, and art—she suggests that the role of science is to explain and predict the world around us—it "converts" information into knowledge. The role of engineering is to apply

E. Bellos, *Sustainability Thinking*, https://doi.org/10.1007/978-3-032-11244-6_7

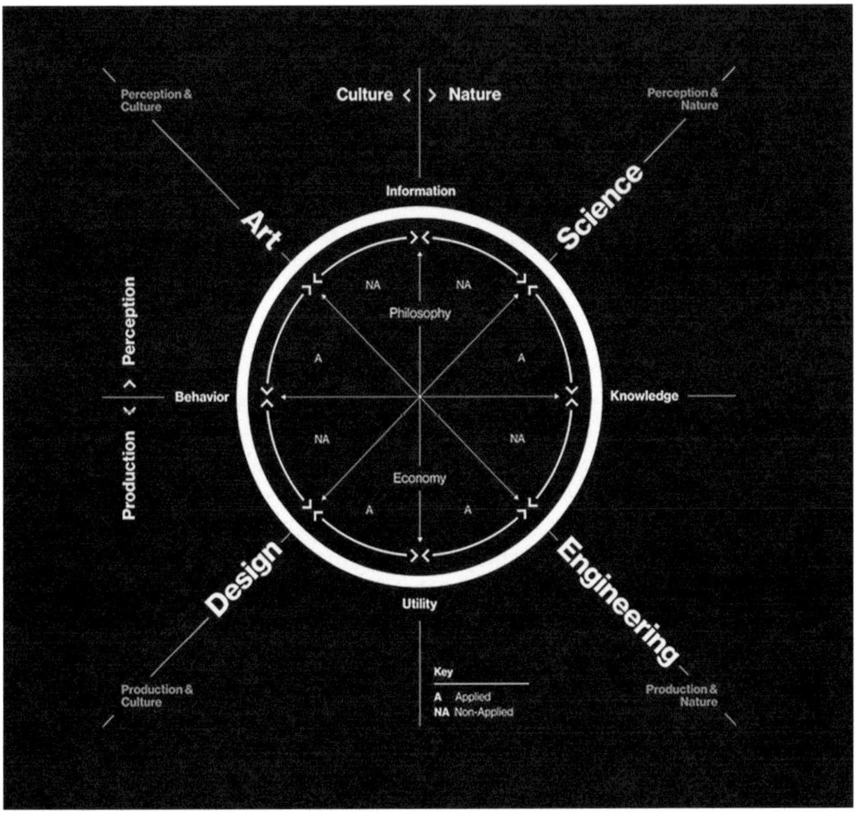

Fig. 7.1 Elements of disciplinary knowledge metabolism. (Source: [189])

scientific knowledge to the development of solutions for empirical problems—it "converts" knowledge into utility. The role of design is to produce embodiments of solutions that maximise purpose and augment human experience—it "converts" utility into behaviour. The role of art is to question human behaviour and create awareness of the world around us—it "converts" behaviour into new perceptions of information, re-presenting the observations that initiated the Krebs-like cycle of creativity in science in the first place. In short, there is repetition, continuity, and change—some content is generated, other content is consumed, some is released and new content is formed [189]. Such fluid knowledge-creating processes and their description are essential to help us understand how learning contributes to the sustainability of the planet and its peoples.

Recent research points to the importance of education for sustainable development. While an educated workforce may contribute to a growing middle class thereby helping to reduce inequalities, skilled labour can further contribute to the faster adoption and diffusion of new technologies that are beneficial to modernisation and development. Increased knowledge and related behavioural changes concerning health risks, for example, can bolster sustainable development in its different dimensions [190]. According to the United Nations, Global Citizenship Education and Education for Sustainable Development can cultivate respect for all, build a sense of belonging to our common humanity, foster responsibility for our shared Earth, and help learners become responsible and active global citizens, as well as proactive contributors to a more peaceful, tolerant, inclusive, secure, and sustainable world. These two initiatives, while respecting cultural diversity, aim to empower learners of all ages to face and help resolve local and global challenges, to take informed decisions and actions for environmental integrity, economic viability, and a just society for present and future generations. SDG 4 concerning inclusive and equitable quality education and lifelong learning opportunities is operationalised by the following target: "by 2030, ensure that all learners acquire the knowledge and skills needed to promote sustainable development, including, among others, through education for sustainable development and sustainable lifestyles, human rights, gender equality, promotion of a culture of peace and non-violence, global citizenship and appreciation of cultural diversity and of culture's contribution to sustainable development". Progress in achieving the stated target is to be measured by the extent to which global citizenship education and education for sustainable development are mainstreamed in national education policies, curricula, teacher education, and student assessment [191].

Citizen Learning

The unexamined life is not worth living—Socrates

Geoff Mulgan in his book *Big Mind* recommends that the citizen learner should be cognizant of three loops of learning in the following ascending hierarchy: (1) adapting thought and action within an existing framework; (2) creating new categories and models to think with; and (3) rethinking how to think. *First-loop learning* involves the application of thinking methods to definable questions as we try to deconstruct, calculate, analyse, and process information using heuristics or frameworks. We begin with models of how the world works and of thinking itself and then gather evidence about our internal and external worlds based on predefined categories. We observe and act when the world does or does not react as expected and adjust our actions accordingly, as well as refine the details of our models in response to the evidence. The reader will suspect at this point that we are talking about *Bayesian learning*. This kind of learning helps us to get by most of the time. It is functional, useful, practical, and relatively easy to deploy. The combination of facts and models enables life to function smoothly and is moreover how our brains work most of the time. When taking into account social complexity it also becomes important to monitor our state of thought such that we move towards *confident humility* as shown in the bottom right cell of Fig. 7.2. Here we have enough doubt

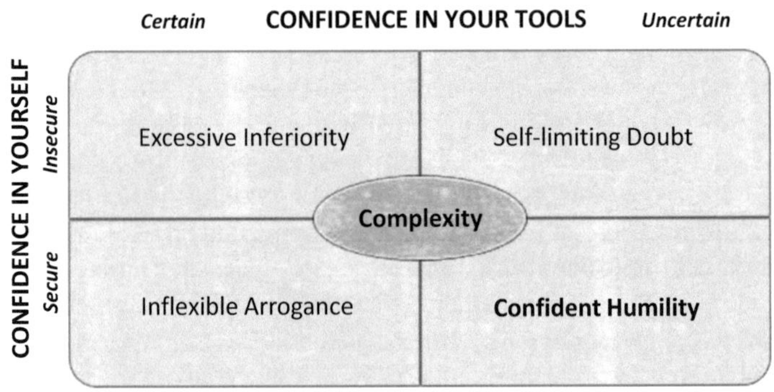

Fig. 7.2 Confidence/over confidence for thoughtful action. (Author adapted from [19, p. 47])

to re-examine our old knowledge but also enough confidence to pursue new insights [19, p. 47].

Second-loop learning becomes necessary when the facts no longer fit the existing models, or when there are too many surprises. It may become necessary to create new categories and models. This is often what we mean by creativity: seeing in new ways, identifying patterns, and generating new frames. There is a trade-off between exploitation and exploration. Exploitation is taking advantage of what we know already works. But because exploitation is so essential to learning, a surprising finding of research on decision-making is that people who are not always consistent often perform better than their peers who are more consistent. *Third-loop learning* involves the ability to reflect and re-reflect on and change how we think—our epistemologies, ontologies, and types of logic. We recognise third-loop learning to have happened when a radically new way of thinking is normalised with its own methods and tools, and its own perspective of what it is that really matters. The psychological growth of any individual involves passing through all three loops repeatedly. Being adept at all three loops helps the individual or organisation to cope with multiple types of thought and the basic characteristics of learning, namely, those that are: iterative, error correcting, tolerant of surprises, and flowing hierarchically from small to large [24, pp. 70–75].

Education for Sustainable Development is transversal across all SDGs and loops of learning. It is a key element of quality education by virtue of its cross-cutting competencies in cognitive, socio-emotional, and behavioural dimensions of learning in all areas of education. More specifically, it places emphasis on competencies related to empathy, solidarity, and action-taking that can help not only for the success of individuals but also for the collective survival and prosperity of the global community. Education for Sustainable Development contributes to moving the global education agenda from an exclusive focus on quality and access, measured mostly in terms of learning outcomes, towards an increased emphasis on how learning contributes to the sustainability of the planet and its peoples. SDG Target 4.7 states that "by 2030, ensure that all learners acquire the knowledge and skills needed to promote sustainable development, including, among others, through education for sustainable development and sustainable lifestyles, human rights, gender equality, promotion of a culture of peace and non-violence, global citizenship and appreciation of cultural diversity and of culture's contribution to sustainable development". UNESCO is the lead UN Agency with the mandate to promote this target. To this end, it has published *Education for Sustainable Development: A Roadmap* (the Education 2030 Framework for Action) which sets out implementation guidance [192]. Monitoring progress reflects characteristics of country education systems, as reported by government officials, and measures what governments intend rather than what is implemented in practice. For each component, a score is calculated by combining a number of criteria to give a single score from zero to one [193]. Environmental education can help making the leap towards the transcendent to give ecological ethics its deepest meaning. It needs educators capable of developing an ethics of ecology, and through effective pedagogy, helping people to grow in solidarity, responsibility, and compassionate care [194, pp. 70,104,116,135,154]. Despite the reality that there are large swathes of the globe for which no data are available, it is nonetheless encouraging that biogeographically rich countries such as Brazil and the Democratic Republic of the Congo score high on this important transversal measure.

The Learning You/Us

Education is the kindling of a flame, not the filling of a vessel—Socrates

First-loop learning may be thought of as curiosity-driven learning. A good place to start is to check whether your initial understanding of relevant terminology, say, "ecology", fits in with the generally accepted definition, or with its major variations and allied terms. A particularly relevant term is "active learning" which is an umbrella name which covers several educational techniques that are, as the name implies, learner-driven. The *Wikipedia* article "Controlled Vocabulary" addresses the problem of synonyms faced by librarians and information scientists to standardise terms for subject indexes. From the middle-ages through most of the twentieth century relevant authorities have tried to establish classification systems to put human knowledge in orderly categories. Melvil Dewey, the professional librarian who developed the time-tested file card and who established the numerical system that bears his name, created a knowledge classification system that remains remarkably serviceable even today. The so-called federated search engines, available at libraries, index hundreds of thousands of electronic journals and many databases. But *how* should such a *first* be undertaken by the non-librarian? This brings us back to Meno's paradox, at least to the point of having to learn that there is now such a thing as general electronic search capability. "Search" is one of the most useful skills of all, yet most educational systems around the world leave it to be organically learned through experience. Edward Tenner suggests that we need to introduce "skilled search" by "desirable difficulty" that schools can teach, for example, by encouraging *Wikipedia* editing. If learners and others were more aware of the limits of their knowledge, they might search more deliberately and carefully. Desired difficulty promotes learning because the brain deems the information in question to be important enough to be retained in long-term memory due to the extra effort involved in acquiring it [71, pp. 109–110]. *Wikipedia* is a partial solution to search results that are distorted by systematic manipulation of search algorithms due to its collective authorship and public editing model which allows for near instantaneous correction. This is particularly helpful for answering questions regarding technical standards and specifications. The curious and motivated learner can build on his or her initial search skills to reach ever greater levels of proficiency. Tenner suggests appending a known author to the subject search which will result not only in a list of peer-referred articles citing the chosen scholar but also reviews and essays in high-quality journals, newspapers, magazines, as well as non-profit and government reports that are not peer-reviewed but are nonetheless important for active democratic citizenship. While few people will need or want to follow through on an entire reading list, a skilled search can map a subject and its major experts and perspectives in a reasonable time together with brief summaries [71, pp. 51, 120–125]. And more help with the *"how"* of search techniques is available from *Google*[1] which is akin to getting advice from Meno himself. Knowing *where* to look to build on an initial skilled search would be an important further step.

After addressing the challenge of "controlled vocabulary" and to advance the selection of knowledge areas that have been brought into sharper focus, Table 7.1 displays seven indicative sustainability disciplines together with their functional elements expressed in the shorthand of equations and the desired outcomes.

Chances are that each of your chosen subjects is covered by a multitude of organisations, platforms, publications, databases, repositories, computer programmes, technologies, as well as networks of other people. Learning that is centred on active research may be thought of as a cone of attention from more divergent sources to more convergent sources. *CiteSpace* is a free software programme that displays patterns and trends in co-cited scientific literature. It sources information from databases, such as the Web of Science, and visually maps the results into network clusters together with the critical points in the development of an academic field.[2] Closer to our interests, Robert Horn maintains a website[3] that bills itself as being visual analytics for public policy, with its creator maintaining that seeing the issues by "mess maps" generates instant stakeholder involvement in contrast to the typical grey literature report of about sixty pages. Similarly, the IPCC has produced a colourful "before and after" picture of climate resilient development centred on the impact of human activities on ecosystems [201, p. 92]. And as noted earlier, if data can also be visually displayed by illustrations, maps, or other means, including interactive methods, the educational value of a presentation or report can be greatly

[1] https://www.google.com/insidesearch/searcheducation/

[2] http://cluster.ischool.drexel.edu/~cchen/citespace/download/

[3] Stanford.edu/~rhorn/

Table 7.1 Seven earth system and sustainability-related academic disciplines

Academic discipline/area	Desired outcome	Purpose and functional elements
Ethics/decision sciences	Just global public goods	Dynamic epistemic equilibria = dual divergent/convergent decision-making and decision-taking + citizen and leader/expert perspectives + discrete assigned roles + context-driven decision-making and decision-taking techniques
Political science/earth system governance	Self-help democracy	Open epistemic democracy = open democracy + epistemic democracy and author [49]
Ecology/Panarchy	Biosphere sustainability	Adaptive cycles = exploitation + conservation + release + reorganisation [195]
Economics	Socially and environmentally responsive economics	Doughnut economics = social foundation + safe and just space + ecological ceiling [196]
Risk management/ergodicity	Holistic risk management	Hierarchical holographic modelling = multi-view image of a system + risk identification at many levels + risk communication of diverse system attributes [197]
Public administration/monitoring and evaluation	Efficient and effective service delivery	Adaptive programmes = adaptive theories of change + adaptive theories of action + adaptive theories of measurement + adaptive risk management + adaptive financing + outlier monitoring + evaluative learning [103]
Design/design abduction	Design thinking	Outcome = What + How [198]
Social psychology	Unifying narratives of hope	Storytelling = character + predicament + attempted extrication [199]
Psychology	Personal mental health and community agency	Happiness = your biological set point + conditions of your life + the voluntary activities that you do [200]

enhanced. To this end, Lenger and Eppler's "periodic table of visualisation methods" modelled on the periodic table of elements offers a number of interesting format options indexed according to structure or process elements, as well as those involving divergent or convergent thinking [202]. Relevant observations, especially if undertaken in real or near real time, can be extra useful, as in the case of wildlife monitoring.

Doing research is easier than ever. Now ChatGPT, DeepSeek, Claude, Google Scholar, Wikipedia, and many other electronic platforms offer a variety of options to become informed on just about any subject that you might think of. *WolframAlpha* serves as a computational knowledge engine across many fields and builds on the achievements of science and other systematisations of knowledge to provide definitive answers to factual queries to the broadest possible range of people—spanning all professions and education levels.[4] The *International Network for Governmental*

[4] https://www.wolframalpha.com/

Science Advice is a collaborative platform of science advisory organisations and national systems that aims to enhance the global knowledge-to-policy interface to improve evidence-informed policy formation across governance levels.[5] The *Stanford Encyclopedia of Philosophy* is an extensive resource of scholarly articles on philosophy from both classical and modern thinkers.[6] The *Online Course Report* website, which caters to working adults, lists the most popular courses by several categories, including for teachers.[7] The Directory of Open Access Journals is an independent database of over 17,000 peer-reviewed open access journals covering science, technology, medicine, social sciences, arts, and humanities, including the journal *Global Sustainability*.[8] The database is supported financially by many libraries, publishers, and organisations to maintain the infrastructure that is needed. The Directory of Open Access Books is an equivalent resource with regard to books. *Nature Briefing* is a service from the Nature group of publications with a wide range of topics. Another useful resource is *frontiersin.org* which is an open-source and open-access platform that covers more than 900 academic disciplines via peer-reviewed articles (mostly on the technical side). A convenient way to stay current with a chosen research topic is to subscribe to reputable newsletters. *Substack.com* is a hosted publishing platform with free and fee paid content that also offers the option to create one's own newsletter.

Steven Pinker suggests that educational institutions from elementary schools to universities should make statistical and critical thinking a greater part of their offerings. Just as literacy and numeracy are considered to be essential building blocks in schooling, the tools of logic, probability, and causal inference permeate every kind of formal knowledge [84, p. 314]. We can add here Massive Open Online Courses (MOOCS) and also readily include knowledge of, and countermeasures to, psychological biases and fallacies as discussed earlier. Pending better curriculum development on wisdom, character education to refine students' moral aspirations can include a wider range of examples from lived experiences thereby allowing for effective shifts in metacognitive dispositions [6]. Several cases of the *bystander effect* and its negative consequences are presented in the book of the same name [203]. One can even attempt to reach *phronesis* by taking *WikiUniversity*'s course on applied wisdom.[9] Moving to more hands-on learning modelled on traditional apprenticeships, Alan Collins and his co-authors have identified four features that can be adapted to learning and knowledge work: (1) *modelling*, that is, demonstrating the task while explaining it aloud; (2) *scaffolding*, that is, structuring an opportunity for the learner to try the task himself; (3) *fading*, that is, gradually withdrawing guidance as the learner becomes more proficient; and (4) *coaching*, that is, helping the learner through some of the challenges along the way [13, p. 164]. And although

[5] https://ingsa.org/about/

[6] https://plato.stanford.edu/

[7] https://www.onlinecoursereport.com/

[8] https://doaj.org/

[9] via https://en.wikiversity.org/wiki/Wisdom/Curriculum

it would be an exaggeration to say that money is no object, several MOOCS offer a variety of free courses and access to learning materials. The *Khan Academy* would be a good foundational educational source before proceeding to more challenging (some free) courses at *FutureLearn*, *Coursera*, *Udacity*, and *EdX*, including on how to learn online. Other options include the Open Courseware Consortium[10] and The Open University.[11] Several online courses are eligible to be used as credits in more traditional university courses.

Initial curiosity-driven research and knowledge acquisition should be developed further with the addition of credible data as the evidence-based movement advocates. Such data may be primary, for example, as generated by public institutions, non-profits, and citizen scientists; secondary from the statistical services of individual countries (which in turn depend on the cooperation of domestic organisations and companies), or tertiary as in the case of data aggregators like multilateral organisations. There are several such sources of *big enough data* that cover a wide variety and numerous subjects. *UNdata* is comprised of 32 databases including for SDG indicators[12]; the *World Resources Institute* compiles data on cities, climate, energy, food, forests, water, and the ocean[13]; The *Global Biodiversity Information Facility* is an international network funded by the world's governments aimed at providing open access to data about all types of life on Earth[14]; the *World Database on Protected Areas* maintained by UNEP is updated monthly with submissions from governments, non-governmental organisations, landowners, and indigenous communities[15]; *The Paleobiology Database* is a public database maintained by an international non-governmental group of palaeontologists that allows for filtering by fossil occurrences according to time, space, taxonomy, and modern and paleogeographic locations[16]; the International Mineralogical Association maintains a growing database of 6006 minerals[17]; The *World Values Survey* analyses people's values, beliefs, and norms in a comparative, cross-national. and over-time perspectives covering political science, economics, social psychology, and other topics[18]; *The World Bank* is a comprehensive, authoritative source that provides global development data about economies and poverty across the world[19]; *Gapminder* is a data aggregator augmented by educational resources[20]; *Our World in Data* provides global data, articles, and charts, including data aggregation across all 17 SDGs via its SDG

[10] www.ocwconsortium.org

[11] www.open.ac.uk

[12] https://data.un.org/

[13] https://datasets.wri.org/

[14] https://www.gbif.org/

[15] https://www.unep.org/resources/factsheet/protected-planet-and-world-database-protected-areas

[16] https://paleobiodb.org/#/

[17] https://rruff.info/ima/

[18] www.worldvaluessurvey.org

[19] https://databank.worldbank.org/home.aspx

[20] https://www.gapminder.org/

Tracker facility.[21] All these data sources are typically free, available in different formats, and variously organised by country, region, or popular analytical units. The *World Futures Studies Federation* offers many types of resources that can be used for scenario development[22] and the University of Denver's *Center for International Futures* has a free modelling system to explore global alternative futures via integrated databases.[23] The National Oceanic and Atmospheric Administration offers many excellent ways and means to support climate analysis and education via its website,[24] while the Fifth National Climate Assessment is the US Government's 2023 report on climate change impacts, risks, and responses.[25] And professional help might be available in the form of your local librarian for materials that are of special interest to you. Add more traditional means of transmitting knowledge, like universities and institutions of higher learning, as well as an assortment of training and apprenticeships, and a rich basket of opportunities becomes available for the curious and motivated learner.

Second-loop learning moves us from first-loop exploitation to second-loop exploration. The hallmark of second-loop learning is creativity. Neuroscientists and psychologists distinguish between crystallised intelligence, based on stored (tacit/intuitive/instinctive) knowledge, and fluid intelligence, the capability to solve novel problems. Individual intelligence is correlated with good performance in all domains, such as higher academic achievement and job performance [82, pp. 228–229]. The characteristics from which creative thinking springs are set out in Box 7.1. Notice that reading and writing (item 9) are key elements that straddle the external and internal worlds of the decision-maker, and like quantitative knowledge (item 8), create the knowledge base on which the other seven cognitive processes can interact. While partially distinct, all nine of these broad cognitive abilities are positively correlated with each other, as well as to varying degrees, with a more global cognitive ability factor. For example, across four studies, *divergent thinking* was found to be a strong predictor of creative achievement in scientific discovery and inventions [88]. A more memorable story of the creative process has been told by Arthur Koestler through his "Creative Trinity" metaphor comprised of the following: the HAHA reaction, the AH reaction, and the AHA reaction. All three domains are characterised by smooth emotional transitions from one to the other. The HAHA reaction is paradox stated; the AH reaction is the absence of paradox; and the AHA reaction is paradox resolved. The HAHA reaction signals the collision of two planes; the AH reaction their juxtaposition, and their AHA reaction their fusion. The three domains of creativity make up a continuum. There is no sharp break where HAHA witticism turns into wit; there is no fixed boundary between the AH and the AHA reactions. There is nowhere a sharp break where art stops and

[21] https://ourworldindata.org

[22] https://wfsf.org

[23] htts://korbel.du.edu/pardee/international-futures-platform

[24] www.climate.gov

[25] http://nca2023.globalchange.gov/

science begins whether we consider art history or architecture, or the culinary arts and psychiatry [10, pp. 184–196].

Box 7.1 The Biology and Characteristics of Cognition and Creative Thinking

The Cattell-Horn-Carroll Theory of Cognitive Abilities

1. *Fluid Reasoning*: "the deliberate but flexible control of attention to solve novel 'on the spot' problems that cannot be performed by relying exclusively on previously learned habits, schemas, and scripts".
2. *Crystallised Intelligence:* "depth and breadth of knowledge and skills that are valued by one's culture";
3. *Short-Term Memory:* "the ability to encode, maintain, and manipulate information in one's immediate awareness".
4. *Long-Term Storage and Retrieval:* "the ability to store, consolidate, and retrieve information over period of time measures in minutes, hours, days and years".
5. *Visual Processing:* "the ability to make use of simulated mental imagery (often in conjunction with currently perceived images) to solve problems".
6. *Auditory Processing:* "the ability to detect and process meaningful nonverbal information in sound".
7. *Processing Speed:* "the ability to perform simple repetitive cognitive tasks quickly and fluently".
8. *Quantitative Knowledge:* "depth and breadth of knowledge related to mathematics".
9. *Reading and Writing:* "depth and breadth of knowledge and skills related to written language".

Source: [88].

Imagination and creativity are central to adapting, creating, and using *active learning,* which is a set of flexible educational techniques that explore learning itself. This wide variety of methods commonly involves learner participation, discussion, reflection, and group work. The following seven engagement techniques are the most consistent with our schema: (1) *background knowledge probes* revolve around asking 2–3 open-ended questions to establish the current understanding of a topic; (2) *analytic teams* consist of learner groups with specific roles, such as "proponent", with each group discussing a chosen topic. Each group member then presents to their constituent assembly their role-based views on the topic; (3) *role play* is similar to the analytic team approach but with more complex roles, such as "policymaker", within a scenario of multiple stakeholders and perspectives. Participants respond to a real problem and propose the collaborative decision that needs to be made. A summary is then presented to the entire assembly to extract lessons learned;

(4) *case studies* create an in-depth, real-world scenario that learners must research and analyse as an example of a concept presented by the facilitator. Solutions are then presented from either the perspective of the decision-maker or that of the stakeholder; (5) *directed paraphrasing* involves rephrasing important passages or concepts directed to a specific lay audience; (6) *autobiographical reflections* ask learners to write short essays about their own personal experience with a course-related topic or question. This can surface both biases and insights; and (7) *student generated rubrics* present participants with examples of varying qualities of work so as to generate criteria for measuring and evaluating the quality of work by each criterion [204]. The University of Illinois provides many other types of techniques and learning resources via its *Online Teaching Activity Index*, including some links to relevant software like concept mapping.[26]

Third-loop learning is the hardest to accomplish using second loop as a springboard. Contributing to the challenge is the insufficient number of academic fields and practices that have either an inter- or transdisciplinary outlook, including in economics. This has driven Jon Erickson to recommend that the very soul of the discipline needs to dramatically change. He suggests that academia can play an important role in redesigning disciplines, curricula, and the very purpose of learning to include new systems of knowing that feed into systems of doing. The task at hand is nothing short of a revolution in classrooms, communities, and legislatures across the world, empowered by and connected to the social movements already under way [151, p. 189]. *The Santa Fe Institute* seems to be on this track via its new Emergent Political Economies research theme which seeks to reimagine the core assumptions of economic, political, and social life by asking "how do we change the rules of the game?" The global clean energy transition and climate change are perfect for thinking about how to change the rules of the game so as to achieve a fundamental reorganisation of energy systems, as well as with those dealing with climate adaptation over the next century. This requires addressing vulnerabilities, the challenges of new energy technologies, wealth disparities, and the political will to pay for changes [205]. As Arthur Koestler has noted, the decisive breakthroughs in philosophy, science, or art are successful escapes from blind alleys, from the bondage of mental habits, from orthodoxy and overspecialisation. The method of escape follows the same undoing and then re-doing pattern as in biological evolution as suggested by Walter Garstang's diagram of the evolution of life, shown in Fig. 7.3, which applies equally well to the evolution of ideas.

Beyond the several false starts that do not lead to $\mathbf{Z_1}$ as shown at the bottom of Fig. 7.3, progress is cumulative only in the macroperspective, that is, by means of the entire cone in the up and to the left direction until $\mathbf{A_9}$. Progress from, say, $\mathbf{A_8}$ to $\mathbf{A_9}$ is not direct but indirect via $\mathbf{Z_9}$. In society, true novelties are not arrived at directly from a previous "adult" theory, but from a new seminal idea. We might say that the evolution of progress follows a draw-back-to-leap pattern [10, pp. 166–196]. And how does this relate to our developing decision-making schema? The individual decision-maker is recommended to *draw back critically to leap forward creatively*; for the decision-making group, it should *draw back critically in conversation to*

[26] https://www.uis.edu/ion/resources/oiai

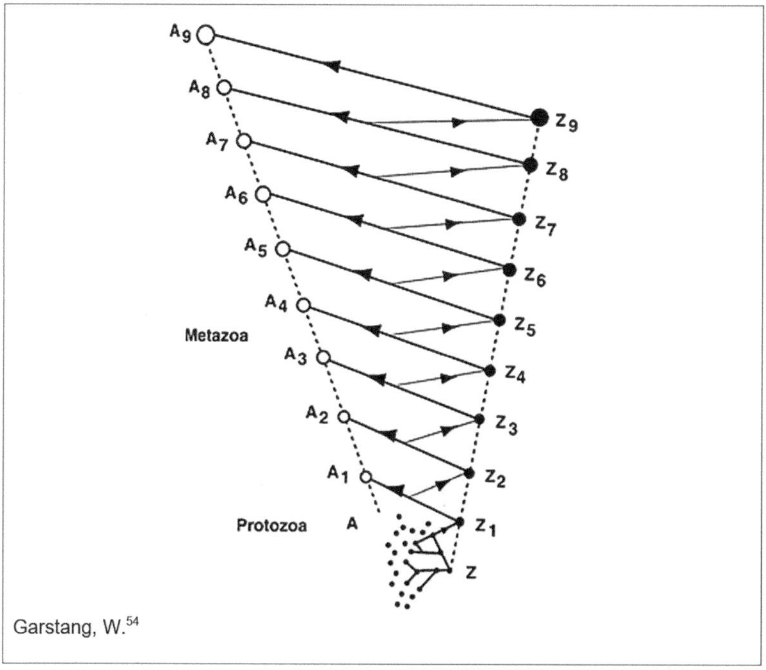

Fig. 7.3 Walter Garstang's diagram of the evolution of life. (Garstang, W. The theory of recapitulation: a critical re-statement of the biogenetic law. *Zoological Journal of the Linnean Society*. 1922 Sep 1;35(232):81–101)

leap forward creatively in dialogue (the different contributions of conversation versus dialogue are discussed in more detail in Chap. 10). Learning, preferably by jumping from one learning loop to a higher one, introduces individual and collective agency to become a directed "learn-critique-create" strategy. Setting resultant aspirational goals prepares the ground for action. Individual creativity cumulates exponentially to become the cycles of societal innovation. Reversals along Garstang's cone can be temporary like when Europeans lost the knowledge for about 700 years to make concrete which the Romans had mastered earlier [28, p. 364], only to be rediscovered and subsequently put to multiple beneficial uses but with some serious unintended consequences like being a major contributor to global warming due to its production methods [206].

Reading for Epistemic Vigilance

We believe scarcely any thing except which pleases us—Blaise Pascal

Naomi Baron, a linguistic professor specialising in reading habits, has found that students often report better comprehension in slower paper-based reading. Print formats, in part because they seem to demand more careful reading, are better suited to the understanding of bigger ideas. Geoff Kaufman and Mary Flanagan found that subjects who saw information on electronic devices were better at low-level recall, but those who saw the same data in hard copy were better able to execute the higher-level *why*, in contrast to the lower-level *how* of a task. It turns out that even the type of font plays a role in the effectiveness of reading. Comparing the performance of undergraduates whose classes had been randomly assigned texts in a variety of fonts, Mueller and Oppenheimer found that those with less legible fonts showed greatest comprehension. Somehow inefficient reading led to more effective learning. The Georgia font seems to be the safest choice because it was commissioned by Microsoft to be equally legible on screen and in laser and inkjet printing [71, pp. 111–115]. And even the placement of writing on a food label concerning calories affects us differently if situated on the left-hand side for English speakers, but on the right-hand side for Hebrew speakers because in both cases the reading direction precedes and in consequence influences our final choices [82, p. 172]. Anne Cutler has found that listeners segment a speech stream based on their cadence of their first language, French into syllables and English by stress placement, which presumably carries over when reading aloud [14]. In everyday life when people place reliance on science when reading and comparing food labels, they rarely make decisions based exclusively on scientific information—they usually also take into account their own knowledge, skills, goals, needs, values, and beliefs [207], as well as tastes!

Research suggests that while skimming is faster than deliberate reading, it also leaves us susceptible to missing important information. Making this type of decision is nearly instantaneous, performed with little or no conscious thought yet a tremendous amount of mental processing occurs during that microcosm of time. Readers often infer the value of a communication from its "envelope", for example, the readily available information surrounding the communication that signals its content. Generally, busy readers attempt to extract as much value as possible with as little time and attention as possible. They don't read line by line, they choose sections that suit their interests, scan others, and skip to what seems relevant. Yet the whole of a text may be greater than its parts. Reading like this takes words out of order and skips over a lot of them completely, which is why skimming is faster than reading, but also why it leaves us susceptible to missing essential information [208]. As consumers of information, we have a role to play in fostering a fuller and more nuanced point of view when reading, for example, by recognising complexity as a signal of credibility. More specifically, we can favour content and sources that present many sides of an issue instead of just one, such as simplifying headlines. We can fight the human tendency to engage in binary comparisons by asking what additional perspectives might be missing between the presented extremes. This applies when we are the ones communicating information as well. New research suggests that when journalists acknowledge the uncertainties around facts on complex issues like climate change, it doesn't undermine the readers trust. Many experiments have shown that when experts express authentic doubt they become more persuasive. When an expert acknowledges uncertainty, it surprises the reader who in consequence pays more attention to the substance of the argument page [19, p. 171]. To stay epistemically vigilant when reading, the following additional practices can help: look for copyright; verify sources by triangulation; assess the credibility of the source; look for a publication date; *Google* the author's expertise with the subject; consider whether the presented information matches your prior knowledge; and consider whether it seems plausible. On the academic front, Retraction Watch[27] basically polices the scientific process and lists retracted articles, including the troubling occurrence of papers that are cited more after retraction than before. Also try to make time to visit FactCheck.org[28] and Politifact.com,[29] which re two reputable online sources to check (mostly) American political claims.

Perhaps any discussion of learning would be incomplete without one's own recommendations regarding what books had the greatest personal influence on him or her. Mine are: *The Quark and the Jaguar* by Murray Gell-Mann; *Antifragile* by Nassim Nicholas Taleb; *The Ghost in the Machine* by Arthur Koestler; *The Master and His Emissary* by Iain McGilchrist; *Gaia: A New Look at Life on Earth* by James Lovelock; *The Dragons of Eden* by Carl Seagan; *Frontiers* by Noel Mostert—and the strongest inspiration for the book you are reading—*A Natural History of Human Thinking* by Michael Tomasello of the Max Planck Institute of Evolutionary Anthropology.[30] And don't let anyone tell you that your favourite books have to be current; any claims that they are dated actually prove that they have stood the test of time. On a personal note, I used to be employed at an organisation where the simple act of reading anything was considered to be actual work. I found this organisational culture foreign to my way of thinking because I always found reading to be informative, or enjoyable, or both.

Learning by Doing

> *Teaching is only demonstrating that it is possible. Learning is making it possible for yourself*—Paulo Coelho

Learning is a predicate to teaching, which can be formally defined as conduct that is primarily or exclusively carried out in the presence of a "naïve" individual or group; brings costs to the teacher or provides no immediate benefit; and facilitates the ability of learners to acquire information or skills more efficiently than would

[27] https://retractionwatch.com

[28] https://www.factcheck.org/

[29] https://www.politifact.com/

[30] Recommended reference books for research methodologies include *The Routledge Handbook of Research Methods for Social-Ecological Systems* edited by Reinette Biggs et al. and *The Power of Survey Design: A User's Guide for Managing Surveys, Interpreting Results, and Influencing Respondents* by Giuseppe Larossi, both of which are free and available online.

otherwise be the case [28, p. 318]. Education is a process involving two sets of participants: teachers who impart knowledge to students, and students who absorb knowledge from teachers. However, as any open-minded teacher will attest to, the arrow of education also works in reverse: it is also about students imparting knowledge to their teachers by asking questions that the teachers hadn't previously thought of, as well as challenging the teachers' original assumptions [140, p. 490]. Beyond the tips offered by Barbara Oakley in her book *Learning How to Learn* and its accompanying online course, the *Feynman Technique* offers a simple yet powerful method for both learning and teaching. It is a study method in five simple steps developed by Nobel Prize winning physicist Richard Feynman for students to learn through the act of teaching: (1) identify the topic you want to learn; (2) explain it as if teaching a child in his or her sixth year of schooling; (3) identify areas of improvement in your explanation; (4) improve and optimise your explanation; and (5) repeat the previous steps until you have achieved mastery of the topic. More specifically, clearly identify the topics and sub-topics you are trying to master; otherwise, you might end-up jumping between topics without developing a deeper understanding of any one of them. This requires you to meet the (hypothetical) child at their level of understanding by using the simplest and most concise language possible. Your explanations need to be short and to the point. To do so, take a critical look at your explanation and try to identify any knowledge gaps where you could not retrieve the relevant information from your memory, together with any areas where you felt that you could not explain the topic in simple terms. Take all the identified knowledge gaps and other areas of improvement and use them to refine your explanation by: developing analogies, keeping a list of terms that are clearly defined before using them in an explanation, and making sure that your explanations are evidence-based and backed by authoritative source material. Finally, mastering the Feynman Technique requires you to take a step back and focus on simplicity and understandability which is difficult to get right the first time. Often, you will need to go through each step of the process at least twice for the method to be effective [209].

Peter Senge in his book *The Fifth Discipline* sets out the potential wisdom of teams that can be attained through *team learning* and collectivising individual intelligence. Most decisions of import are now made in teams, either directly or due to the necessity for teams to convert discrete decisions into action. Team learning is the process of aligning and developing the capacity of a team to create the results its members truly want to achieve. Even though it involves individual skills and areas of understanding, team learning is a collective discipline. When individuals are empowered by virtue of their *alignment*, the necessary conditions are created that can in turn empower their whole team. Individuals do not abandon their personal interests in favour of the larger team vision; rather the shared vision becomes an extension of their personal visions. Team learning revolves around four key factors: First, there is a need to think critically and insightfully about complex problems while devising means to counter any larger or stronger organisational forces that may be threatened by teams more intelligent than a particular member. Second, innovative and coordinated action is necessary involving *operational trust* gained through the mutually complementary actions of other team members garnered through experience. Third, a nested learning team needs to nurture other learning teams by inculcating the practices and skills it has developed given the interdependencies of implementing actions. Mastering the practices of *dialogue* and *discussion*, the two distinct ways that teams communicate are critical to team learning. In dialogue there is a free and creative exploration of complex issues, "deep listening", and the suspension of one's own views. By way of contrast, in discussion different views are presented and defended, while there is a search for the best option to support any decisions that must be made timeously. The discipline of team learning involves developing the capability of distinguishing between dialogue and discussion and moving consciously between them when they are not complementary. Finally, like any discipline, team learning requires practice. This entails convening *dialogue sessions* where learning teams "learn how to learn" converting team learning into a *team skill*. The basic conditions for dialogue sessions include:

- Explaining, with examples, the ground rules of dialogue.
- Enforcing those ground rules so that if someone finds oneself unable to suspend his or her assumptions, for the team to acknowledge retuning to "discussing", not "dialoguing" if this is the case.
- Encouraging team members to raise the most controversial and subtle issues essential to the team's objectives.
- Having all the individuals who need one another to execute a decision successfully act together [145, pp. 216–252].

Counterintuitively, dialogue sessions should begin by abandoning best practices. Best practices imply that ideal routines are already in place, but adopting *process accountability* elevates the search for better practices to an ongoing learning modality. One of these best practices should include establishing *psychological safety*. In such learning environments, participants feel confident that they can question and challenge the status quo without being punished. To this end, psychological safety often starts with leaders role-modelling authentic humility [19, p. 256]. It is also important at this point to note that dialogue is homologous to divergent thinking, whereas discussion is homologous to convergent thinking (when either is used by individuals or groups).

Citizen Science

Our task is not just to train more scientists but also to deepen public understanding of science—Carl Sagan

Although there are multiple and wide-ranging definitions of what citizen science is, the following from the US *Crowdsourcing and Citizen Science Act* (2016) is in line with the citizen philosophy conception more than others: the term citizen science means a form of open collaboration in which individuals or organisations participate in the scientific process in various ways, including enabling the formulation of research questions; creating and refining project design; conducting scientific experiments; collecting and analysing data; interpreting the results of data; developing technologies and applications; making discoveries; and solving problems. The United States federal government supports citizen science in part via its website *citizenscience.gov*, which in turn furnishes a toolkit, including case studies, a document library, guides, and other resources. When coupled with emerging technologies, such as web applications, mobile devices, and smart sensors, citizen science enables the collection and analysis of large volumes of geo-coded data to aid decision-making, educate the broader public, and enhance public participation. While Earth observation is the main source of remotely sensed big data, citizen science enables timely, cost-effective collation of local data across dispersed locations [210]. Based on Amazon's Alexa rankings *iNaturalist* is currently the most popular citizen science website followed, respectively, by *eBird* and then *Zooniverse* [211]. Although I have personally contributed to *Zooniverse* wildlife projects and was satisfied with the experience overall, one way it can be improved is to directly share with contributors the final results of a relevant study in summary form. With appropriate protocols, training, and guidance, citizen scientists can collect data of equal quality to those collected by experts. Consider that *Zooniverse* projects have yielded more than 50 peer-reviewed articles on topics ranging from galaxies to oceans. Data from *eBird* have been used in at least 90 peer-reviewed articles and book chapters covering topics in ornithology, ecology, statistical modelling, and climate change. Supporting initiatives for citizen science include formal data policies; data management and quality control plans, as well as many guides, tools, and templates for project planning, and monitoring and evaluation.[31]

To facilitate further citizen science development, a consortium of individuals and organisations has created the international Citizen Science Association[32] with the aim of promoting citizen science best practices in ethics, project evaluation, scientific rigour, and data management. Citizen science projects could be coordinated around the world to synthesise and analyse their diverse data sets by means of Citizen Science Centres. These could serve to create centralised repositories of volunteer-collected data on public goods such as air and water quality, precipitation, biodiversity, phenology, astronomy, human health—and critically—as coordinating centres in disaster management. Centres could also help to field questions about citizen science data and techniques for achieving educational and community development goals. As such, Citizen Science Centres would be excellent strategic investments for both private and government foundations [212]. Similarly, Louis Liebenberg has proposed a data-centric approach to citizen science whereby sections in scientific journals could be reserved for data and papers provided by citizen scientists. He notes that while this may not be viable for paper editions of journals, electronic editions could provide virtually unlimited space to publish data. This could then serve as a platform for new citizen scientists to get their data published

---

[31] See citizenscience.org, citizensciencealliance.org, and citsci.org

[32] www.citizenscienceassociation.org

and receive critical feedback from their peers and professional scientists thereby building their practical experience [213, p. 225].

More recently, top-down initiatives bringing together a variety of actors have emerged outside institutional frameworks, using online platforms. For example, a community of biologists, engineers, and developers has emerged on the Just One Giant Lab collaborative platform to develop low-cost, open-source solutions against viruses. This platform, which was developed with Leo Blondel (Harvard University) and Thomas Landrain (La Paillasse, PILI) over the past few years, is designed as a virtual, open, and distributed research institute aimed at developing solutions for the SDGs. Communities use it to self-organise and provide innovative solutions to urgent problems requiring fundamentally interdisciplinary skills and knowledge. The platform facilitates coordination by linking needs and resources within the community, animating research programs, and organising challenges [214].

Citizen science also has a lot to offer to policymakers. *iNaturalist* is an online social network of people sharing biodiversity information by mobile phones and other means to help each other and scientists learn about nature. Observations are shared with scientific data repositories like the *Global Biodiversity Information Facility*. It is essentially a crowdsourced species identification system and an organism occurrence recording tool that also helps the user identify the organism they are looking at. In other words, it is bottom-up citizen science enterprise that brings people and nature closer to each other [215]. This initiative can be readily emulated to similarly obtain on-the-ground sustainability data interactively from citizens. Recording rising water levels (seas or floods) from previously known high water marks in local political jurisdictions, any drying local landscapes, and personally experienced extreme weather events are just a few of these possibilities. The problem of appropriate aggregation methods and curation will continue to be a challenge, but a lot has been learned due to the COVID-19 pandemic. Additional bottom-up citizen action can be just a click away. Since its founding in 2009, the *Ecosia* search engine has helped finance reforestation efforts across the world. *Ecosia* uses 47.1% of its income (80% of its profits) to support tree planting projects, such as the "Plant a Billion Trees" programme run by The Nature Conservancy. Other projects have included the funding of reforestation as part of the Great Green Wall project, which aims at preventing desertification in Africa. *Ecosia* is available as a web browser on PCs and MACs, as well as a mobile "app" on Android and iOS devices [216].

Gaming: An Example of a Sustainability Learning Technique

The most important attitude that can be formed is that of desire to go on learning—
John Dewey

Assuming that having a grounding in traditional ecology, population biology, biogeography, systems science, environmental management, economics, public policy, or several other related academic disciplines helps in acquiring the interdisciplinary knowledge necessary to address global sustainability, we will nonetheless need a more focused entry point to achieve hands-on learning. Computer games, especially, the rise of online gaming worlds have created a new and unexplored way to disseminate and undertake global change research. Modern play methods imitate the development and testing of scenario research models whereby players can explore and select hypotheses so as to understand a game's environment. In consequence, players get involved in managing a multitude of options, not unlike researchers mining databases and other scientific sources for evidence that may help in forecasting global futures. Modern gaming is an opportunity for the global change scientific community to capitalise on the creativity of citizen scientists and philosophers in thinking through and designing different scenarios while at the same time creating and receiving useful information [217].

Developed and hosted by MIT's Sloan School of Management, one such free online educational game *World Energy: A Climate and Energy Policy Negotiation Game*[33] is a role-playing simulation where in a group setting (in-person or virtual) participants adopt different roles. Typically, the Climate Action Simulation is played with teams divided into six stakeholder groups: *clean tech, conventional energy, industry and commerce, world governments,* and *climate justice hawks*, but can be variously modified, for example, to include developed, developing, and emerging nations. Participants discuss, negotiate, and test possible strategies and dynamics, such as implementing energy efficiency technologies, carbon pricing, and subsidies to reduce greenhouse gas emissions. Proposals are then checked against science-based research using the game's energy and economics simulator called En-ROADS,[34] which provides participants with science-based feedback on their investment and policy proposals thereby promoting learning without being told what should be done ahead of time. The simulation focuses on energy supply, economic consumption, agriculture, and related issues that can form a part of sustainability courses, or other activities. The game is conducted as an emergency climate summit organised by the United Nations to establish a concrete plan. It is run by a trained facilitator acting as the UN Secretary-General who has convened the meeting and who leads participants throughout the experience. Game and game facilitation resources, as well as a training programme are also freely available through www.climateinteractive.org.[35] Taking part in the game improves participants' knowledge of the actions needed to address climate change, boosts both personal and emotional engagement with climate issues, and leaves participants feeling empowered to address climate change [218].

[33] https://www.climateinteractive.org/climate-action-simulation/

[34] https://en-roads.climateinteractive.org/scenario.html?v=22.5.0

[35] https://www.climateinteractive.org/climate-action-simulation/leading-the-climate-action-simulation/#materials

Critical Thinking

I cannot teach anything to anyone, I can only make them think—Socrates

In July 2022, a natural language *WolframAlpha* search using the term "critical thinking" yielded the following response: "development of this topic is under investigation…". After more than 30 years of development this search result suggests, that as ambitious as this project has been, it has not managed to effectively address a key attribute of human thinking and intelligence. I ascribe this to the inherent ambiguity of language and the complexity of the topic itself rather than some computational or intelligence failure. If anything, this manifest difficulty is another demonstration of the difference between mêtis and metasophia thinking versus techne and episteme—the former not being amenable to formulaic manipulation, the latter more so—all themes of our present undertaking. As is often the case, a detour back to the ancient philosophers holds additional insight into the fundamental workings of critical thinking. Plato, more than 2500 years ago set out the laws of thought comprised of *identity* (whatever is, is), *non-contradiction* (a thing cannot both be and not be), and the *excluded middle* (a thing must either be or not be). And according to Saaty (2014), *comparison* ought to be the fourth law of thought. Without comparison we cannot know "what is, and what is not". The law of comparisons underlies the other three laws of thought. It resides in our biology and enables us to discriminate between being ourselves and sensing that there is something else that is not ourselves. With respect to measurement, given two entities that are homogeneously close, it is possible using expert judgement to estimate the dominance of one over another numerically and derive priorities from such comparisons [219]. Comparison is fundamental to science in experiments and quasi-experiments, for example, when a treatment group is compared to a control group; when establishing indicator baselines so as to compare progress in relation to starting states; when comparing scenarios or simulations to a base case in forecasting; when competing theories are assessed, and in many other applications.

Plato's student, Aristotle, lay the groundwork for formal critical thinking and logic in his book *Prior Analytics* wherein he defined *syllogisms* as taking a three-part form: (i) a general proposition, (ii) a minor proposition, and (iii) a conclusion [220]. For example:

(i) All mammals are animals.
(ii) Dogs are animals.
(iii) Therefore, dogs are mammals.

The conclusion is logically true *if* the propositions are true. However, some conclusions can be logically false (paradox), but empirically true (observation).[36] Therefore, compound or complex language statements or propositions should also

[36] For example, the following is logically false: all mammals are herbivores, ostriches are herbivores; therefore, ostriches are mammals. This is false because the major proposition is false leading to a conclusion that can be empirically refuted (only mammals suckle their young).

be subjected to further testing, or compared to empirical evidence, or both. The Greek concept of *aletheia*, meaning "unconcealing", implies a number of facets of truth which are themselves unconcealed. It is a *process* of removing things, rather than putting things together. Importantly, that process is also a part of the truth. This idea of truth as unconcealing contrasts with the notion of truth as correctness, which is something that is static. Truth as unconcealing is progress towards an unknown, whereas truth as correctness can in principle be fully known. In short, truth is process, not object [5, pp. 151, 154].

Modern versions of critical thinking, and the use of the term itself, can be traced to the American philosopher John Dewey (1910) who outlined five of its phases:

(i) *Suggestions*, in which the mind leaps forward to a possible solution.
(ii) An intellectualisation of the difficulty or perplexity into a *problem* to be solved, a question for which the answer must be sought.
(iii) The use of one suggestion after another as a leading idea, or *hypothesis*, to initiate and guide observation and other operations in collection of factual material.
(iv) The mental elaboration of the idea or supposition as an idea or supposition (*reasoning*, in the sense on which reasoning is a part, not the whole, of inference).
(v) Testing the hypothesis by overt or imaginative action.

Additionally, educators typically understand critical thinking to have at least three features: it is done for the purpose of making up one's mind about what to believe or do; the person engaging in the thinking is trying to fulfil standards of adequacy and accuracy appropriate to the thinking; and this exercise fulfils these standards to some threshold level. In essence, these three features constitute *careful goal-directed thinking*. The opposite side of the coin also helps to select by elimination the key components of critical thinking: it excludes jumping immediately to conclusions; excludes not suspending judgement no matter how strong the evidence; excludes unquestioned thinking from an ideological or religious perspective; and excludes using an algorithm routinely to answer a question. For our purposes, we will adopt the conception of critical thinking, which is grounded in rationality, more specifically, epistemic rationality (fitting one's beliefs to the world), and instrumental rationality (optimising goal fulfilment).[37] Within this rubric some of the most important critical thinking abilities can be summarised as follows: *observational abilities* require an understanding of the difference between observation and inferential conclusions. These amount to being able to articulate, including measure, what one has observed through one's senses. *Questioning abilities* require an understanding of the concepts of ambiguity and vagueness. These are the abilities that can transform an inchoate sense of perplexity into clear question(s) and are therefore critical to formulating research and evaluation questions. *Argument*

[37] Credit: Stanovich, Keith and Paula Stanovich (2010), *A Framework for Critical Thinking, Rational Thinking, and Intelligence.*

analysis abilities require an understanding of the concepts of argument, premise, assumption, conclusion, and counter-consideration. *Inferential abilities* require an understanding of the difference between deduction and induction, as well as that of necessary versus sufficient conditions. They revolve around judging and deciding skills. More specifically, they involve the ability to translate natural-language statements into statements using the standard logical operators (see truth table below), to comprehend argument forms and those containing symbols, to determine whether an argument's form and conclusion follows necessarily from its premises, and to reason with logically complex propositions. *Experimenting abilities* require understanding of the concepts of hypothesis, null hypothesis, assumption, prediction, appraisal of evidence, and statistical significance and its correct use. Knowing how to design and execute an experiment is not only important in scientific research, it is also necessary for everyday life. Scientific experimentation skills require knowledge of the difference between an experiment and an observational study, and in particular of the difference between a randomised controlled trial, a prospective correlational study, and a retrospective study. *Imaginative abilities* are those that are directed at finding the correct causal explanation of a general phenomenon or particular event and to imagine possible contributory mechanisms. Exploring what a policy or plan of action to take requires the generation of options and consideration of their possible consequences, including their risk profiles and any unintended consequences. Domain knowledge is required for such a creative activity. *Emotional abilities* contribute to the critical thinking process by reinforcing curiosity and resolved puzzlement. These abilities gain importance in collective settings by enabling the recognition of one's own and others' emotional commitments and reactions. Several tests[38] are available to gauge critical thinking abilities, the most cited being the *Watson Glaser Critical Thinking Appraisal* [221].

Thinking Critically, Collectively

Each of us has a motive to prefer our truth, but together we're better off with the truth—
Steven Pinker

It is self-evident that critical thinking is primarily the province of the individual, but it too can and should be collectivised. Although there are multiple approaches and techniques that improve collective critical thinking, two are particularly suitable to "book-end" the process: "red teaming" prior to decision-making, and summative evaluation after decision-taking, with all lessons learned fed back into any existing formal planning cycles of organisations. The entire concept of red teaming is based on the assumption that institutions are typically ineffective at assessing themselves thereby leaving them vulnerable to dangerous blind spots that adversaries can

[38] *Cornell Critical Thinking Test, California Critical Thinking Skills Test,* and the *Collegiate Learning Assessment.*

exploit. Red teaming is a structured process that seeks to better understand the intentions, interests, and capabilities of an institution through simulations, vulnerability probes, and alternative analyses. Although the approach is underexplored by those facing threats, complex decisions, and strategic surprises, red team methods have been used in a wide range of fields and tailored to various needs. By employing a red team, institutions can get a fresh and alternative perspective on how they do things, test unstated assumptions, identify blind spots, and promote improved performance. In red team exercises, opposition forces are like vaccinations, they introduce foreign substances into the body that attack weaknesses. Red team exercises, not unlike vaccinations, aim to strengthen the body's response to future risks. Senior government officials and other decision-makers are increasingly unable to process complex information in a limited timeframe because of the many factors and actors that need to be taken into account. For these reasons, the use of external facilitators is the most effective approach provided they remain distant enough from day-to-day operations so as not to be similarly biased as their target audience, but close enough to know the context, especially the "centre of gravity" of an organisation whose failure will cause the most damage. From long form workshops, to shorter, sometimes even anonymous exercises, red team techniques aim to find all the vulnerability gaps in an organisation's plan before they have a chance to become hazards. And over time, an iterative red teaming process can help expose different vulnerabilities at different, less critical parts of an organisation [112, pp. 275–277]. In his book *Red Team,* Micah Zenko provides sources and methods together with six recommended best practices: (1) first, the boss must buy-in; (2) be outside and objective, while inside and aware; (3) be a fearless sceptic with finesse; (4) have a big "bag of tricks"; (5) be willing to hear the bad news and to act on it; and (6) "red team" just enough, but no more [87, pp. 235–236].

Evaluation, especially programme evaluation, is a much better-established discipline than red teaming. Although evaluations can be undertaken at any stage of an intervention, summative evaluations at the end of a project or programme are much more common. Programme logic is often used to describe the way in which an implementation programme fits together, usually in a simple ascending sequence of *inputs*, *activities*, *outputs*, *outcomes*, and *impacts*, collectively known as a results chain. Programme theory goes a step further and offers an explanatory account of how the programme works, with whom, and under what circumstances. Programme theory may be viewed as an elaborated programme logic model, where the emphasis is on causal explanation using the idea of "mechanisms" that are at work with other conditions remaining the same in one temporal and one spatial scale generally assumed to obtain. Monitoring the effects of implementation, particularly if linked to early warning and ex-post evaluations is generic components of risk management. Yet it is uncommon to see risk indicators incorporated in a results chain, either as adjuncts to more traditional goal-oriented ones, or as sign-posts to impending breaks in the results chain. Risk indicators designed for this purpose in the form of "jumping indicators" can fill this gap in both states of equilibrium and transition [103]. Governments and institutions have tens of thousands of evaluations that were commissioned to assess whether a policy, programme, or project was completed, what impediments materialised, whether it worked, and how much it cost. UNDP alone has more than 6600 such studies. Yet transparency is not the norm, some are published in academic journals, most are not. Thousands of evaluation studies are buried in institutional vaults even though rigorous, independent evaluations offer a robust way for scientists, policymakers, and citizens to learn which policies should be modified or expanded and which should be terminated [159].

The six most important conditions for successful evaluations are: (1) predetermined and transparent evaluation criteria that are appropriately applied in the execution of the evaluation; (2) clear guiding questions and sub-questions for the evaluation to answer; (3) availability of relevant data and information; (4) suitable research and analysis methodologies that are appropriately applied; (5) insightful and actionable recommendations for the attention of relevant stakeholders; and (6) adequate funding. There several resources on the Internet to assist with designing, commissioning, or undertaking evaluations, including The World Bank's *Nuts and Bolts of M&E Systems.*[39]

In the age of AI, thinking critically and collectively has become even more important. What would it mean for us to live in a world where scientific theories, political manifestos, and even religiosity are shaped by a non-human intelligence that knows how to exploit with unsurpassed efficiency the weaknesses, biases, and addictions of the human mind? [76, p. 208]. In a paper titled *The Illusion of Thinking* put out by Google in 2025 the authors reach the conclusion that fundamental limitations in Large Reasoning Models exist since these models fail to develop generalisable reasoning capabilities beyond certain complexity thresholds despite sophisticated "self-reflection" mechanisms. Particularly concerning is the counter-intuitive reduction in reasoning attempts as problems approach critical complexity pointing to an inherent scaling limit. In simpler problems, reasoning models often identify correct solutions early but then inefficiently continue exploring incorrect alternatives—an "overthinking" problem. At intermediate problem complexity, correct solutions arise only after extensive exploration of incorrect paths, and beyond a certain complexity threshold, models completely fail to find correct solutions. By way of comparison, Standard Large Language Models outperform Large Reasoning Models at low complexity, but Large Reasoning Models excel at moderate complexity, and both collapse at high complexity [222].

Public Issues Delphi and Delphi-like Participation

> *If you want to teach people a new way of thinking, don't bother trying to teach them. Instead, give them a tool, the use of which will lead to new ways of thinking*—
> Buckminster Fuller

The Delphi technique, named after the ancient Greek oracle, is a social science methodology that began in the 1960s and has been in continuous use since then with many innovative modifications along the way. It belongs to a class of techniques

[39] https://www.worldbank.org/en/topic/poverty/publication/nuts-and-bolts-engendering-monitoring-and-evaluation

known as participatory methods and has proved to be useful especially in futures research and scenario development. One might imagine that explorations of the future through public participation began about three million years ago in Africa as groups of humans clustered together gazing at the evening sky contemplating their respective fortunes. These informal group meetings evolved into the "palavers" common in much of Africa today, and the less common, but larger recent Citizen Conferences of Paris. The usefulness of participatory methods revolves around the extent to which the recipients of a decision are involved in making the decision itself and thereby contributing to its public acceptance. And the extent to which the decision-makers themselves are involved in the corresponding citizen process contributes to the extent that decision outcomes will be implemented with ease and speed. The greater the scope of alternative futures considered in the process, the more likely it is that the decision outcomes will have a positive and lasting impact. The Internet and other forms of modern communication have dramatically increased public participation in the policy process and broadened interaction globally. The results of these processes tend to be more normative (what the future should be) rather than analytic (what the future might be, based on the empirical conditions of the past and present). Participatory methods, such as Delphi, can identify the aspirations of a community and general strategies to achieve them [223] and set the stage for follow-up analysis and plans. The main strength of Delphi is its ability to explore systematically issues that require judgement—more specifically the desirability of some future state. This is done by posing questions which may involve a moral, political, or social dimension quite distinct from the disciplinary expertise involved in judging likelihood. At one extreme are questions about the future for which factual answers exist and thus require minimal judgement, and at the other, are those involving a high degree of uncertainty. The variables identified in the questionnaires can subsequently be used to obtain historical data from the most authoritative sources and subjected to time-series or Trend Impact Analysis to get forecasts for each variable. The criteria for assigning a weight to a variable can be the number of people affected; the magnitude of the effect; whether some groups might be affected differentially; the time over which the effect would be felt; and whether the effect is reversible [224]. The frequency and magnitude of an effect are standard risk assessment criteria.

The Delphi method is a controlled discussion conducted in a safe space provided by the anonymity and transparency of the process. Traditionally, the Delphi technique involved at least two rounds of questions administered in a group setting to elicit informed responses from a panel of selected experts. Identified shortcomings in the development of Delphi included the length of time it took to conduct multiple rounds of questioning together with the challenge of participants dropping out between rounds. This led to the creation of "Real-time Delphi" which was greatly facilitated by electronic means of administration. Furthermore, it was realised that Delphi could be applied more broadly given the advent of the Internet. The essential difference between a regular Delphi and a Public Delphi is that participants select themselves by responding to the public Delphi and know what ideas and judgements are being considered, *while* the Delphi questionnaire is being administered. Statistically representative samples are not necessary and anonymity can still be guaranteed in the public Delphi by not requiring respondents to identify themselves [223, 224]. A recent proposal has suggested that not only final decisions should be taken into account, but also the arguments in favour of the final selected alternatives, especially if positions have shifted across deliberation rounds [225] (which is effectively a borrowed Delphi technique).

In the specific context of sustainability governance, the importance of participation in policymaking and decision-taking has been repeatedly stressed by scholars and practitioners. The inclusion of different actors in the drafting of policy proposals is deemed to yield more effective results on the basis of the aggregation of resources, increased legitimacy, and mutual learning [190]. The many existing and possible Delphi techniques are particularly useful for futures analysis to explore the kind of questions relevant to socio-ecological studies and how research approaches, such as adaptive co-management are co-evolving together with other approaches [226]. The possible addition or refinement of analysis criteria makes the Delphi method a natural fit with another social science methodology, namely, MCDA, which has its own particular variants and procedures. Furthermore, each of these techniques can be combined with other methods to produce tailored decision support information. But even the most recent and sophisticated methods in the literature do not recommend or identify any follow-up periods of review for purposes of re-reflection and needed adjustments, which are the final elements of our decision-making and decision-taking schema presented in Chap. 10.

Truth Tables: An Initial Critical Thinking Technique

If we can't think for ourselves, if we're unwilling to question authority, then we're just putty in the hands of those in power—Carl Sagan

Truth tables work by breaking down relatively complex sentences into their component parts so as to determine the accuracy of a stated proposition as being either true of false. Truth tables for more than four variables are rarely given since even such a modest number of inputs produces 256 truth values given that a function may produce either a true or a false answer for each combination as calculated in the form of a double exponential. For eight variables, the possible values are 1.2×10^{77} [227]. Such astronomical numbers are one reason why deceit and lies can remain invisible requiring additional hierarchical layers of critical thinking for the accurate analysis of extensive and complex communications, such as those found in political speeches and single-source reports. Consider that in the case of COVID-19, contracting the virus despite being vaccinated may have led to differing severities of the disease from mild symptoms up to, and including death, making any simple binary "anti-vax" sentiment suspect. Truth tables get us closer to the truth by simply holding that a conclusion must logically follow from its premise. Consider the following claims of Eric Li (EL), a China area expert:

> (EL): Westerners always assume that multi-party election with universal suffrage is the only source of **political legitimacy**. I was asked once, "The Party wasn't voted in by election. Where is the source of legitimacy?" I said, "How about *competency*?"... Now, most political scientists will tell us that a one-party system is inherently incapable of self-correction...Mao was the father of modern China, yet his prolonged rule led to disastrous mistakes. So the Party instituted <u>term limits</u> with [a] mandatory retirement age of 68 to 70...Now I would venture to suggest [that] the Party is the world's leading *expert* in <u>political reform</u> [228].

We can match the claims from the above quote, respectively, in **bold** or *italics* or <u>underlined</u> and label them as variables "p", "q", and "r". Read in the context provided, it can be inferred that if there is political legitimacy, then there will also be competency and political reform. The results of this combination of propositions via the logical operators are provided in the last column of Table 7.2.

In effect, we are only interested in the top four rows of the truth table where political legitimacy (p) is given as being true, as well as the instances where both competence (q) and political reform represented by term limits (r) are also true so as to test the stated set of claims. This is the case only in row 1. In everyday life, however, we are not limited to just one or two sources of information. We can seek additional information and verification elsewhere, a tendency that makes original propositions either stronger or weaker based on the degree to which they are either supported or refuted by available evidence *as if* they are truth statements. In doing so, however, we would be discounting Plato's law of the "excluded middle" while recognising the existence of an *informal fallacy*; otherwise, viable alternatives would be ignored [229].

In our example, President XI was re-elected to an open-ended term in 2022 by the 5th Communist Party Congress, and in effect, made President for life—contrary to EL's claim that the Party's reform expertise has been demonstrated by the introduction of leadership term limits. This new evidence demands that the variables be

Table 7.2 Truth table on whether political legitimacy implies competency and political reform

p	q	r	$p \Longrightarrow \quad q \wedge r$
T	T	T	T
T	T	F	F
T	F	T	F
T	F	F	F
F	T	T	T
F	T	F	T
F	F	T	T
F	F	F	T

Source: Wolfram Analytics

updated so as to identify in which row of the truth table term limits, and hence political reform, are now false (F), while all other propositions remain true (T). This only occurs in row 2 with a result of F in the last column—proving that the initial set of propositions as laid out by the speaker are *logically* false. Yasheng Huang, another China area specialist, takes an opposing view to EL regarding the source of political legitimacy and claims that "…the single most important benefit of democracy is its ability to tame violence…" [230], in other words, not competence. The violent attack on the U.S. Capitol on 6 January 2021 by an insurrectionist mob unhappy with the Presidential election results followed by similar events in Brazil in 2023 undermines Huang's claim. We can follow a similar analytical procedure as in the case of EL, but we know already that this claim is not accurate, or revolves surreptitiously around what is meant by the word "tame". Democracy is a conversation, and conversations rely on language, and so do large language models. By high-jacking language, computers can make it even more difficult for large numbers of humans to conduct a meaningful public conversation [76, p. 210]. Consider that ChatGPT was asked: "is the single most important benefit of democracy its ability to tame violence?" Its top line reply was "while democracy certainly has the potential to mitigate violence, it would be an oversimplification to say that the single most important benefit of democracy is its ability to tame violence". On one view, this is a confirmatory assessment in noting that the given proposition is an oversimplification, but a kind of circular logic on another view".[40]

Parenthetically but interestingly, ChatGPT uses a hundred times more energy than a Google Search query [231]. When asked how much energy it uses to generate an answer, ChatGPT replied, "I don't have specific energy consumption data for my own operation".

In conclusion, given the vast number of possible statement combinations in political and other discourse, a default intuitive appraisal would be needed as opposed to only a simple true-false judgement. This implies that not only initial assessments may be wrong, but also that any given propositions may be insufficient for the task of verification due to their number, combinations, significance, or effective veracity. In other words, epistemic vigilance and critical thinking would still be needed when evaluating any complex statement or proposition.

[40] ChatGTP 3.5, September 25 version, 1 reiteration.

Chapter 8
Expert Roles

Democracy will never be supplanted by a republic of experts – and that is a very good thing—Thomas Piketty

So far, we have engaged in a coarse-grained discussion of the role of biology, philosophy, science, and technology in human affairs, but we have yet to trace our metaphorical map down to the level that roles play in group decision-making—that lever of wisdom and driver of thoughtful action. Here, in finer-grained mode, we are more concerned with the particular role of "expert" and his or her guiding function in group decision-making. Malcolm Gladwell in his book *Outliers* popularised the notion that 10,000 hours of practice were needed to become an expert in one's chosen field. More recent research shows that expertise is basically pattern recognition stored in long-term memory built up through many repetitions, timely feedback, a conducive environment, as well as deliberate practice [232]. Within a particular practice domain, experts can make themselves into accessible models for imitation by student practitioners by making explicit the categories they use to organise information, which in and of themselves are instructive about the principles that the experts use in their own thinking. Experts can make their knowledge visible by delineating the aggregation steps that were used, exaggerating salient features, and recommending new organising categories. Experts literally have a different outlook about a problem than their apprentices—they take in the big picture more rapidly and completely and shift more easily among visual fields—all while focusing on the most important aspects of a problem. This is different than beginners' tendencies which are widely divergent and idiosyncratic in their search for possible solutions, but research suggests that the ability to coordinate such moments can be acquired by practice. Although experts are not aware of how they engage in seeing, eye-tracking technology can reveal this aspect of their expertise and guide the gaze of beginners with unobtrusive cues about where to look. The common ground established by shared attention becomes especially important for teams working together to solve a problem. In addition, when each member is charged with maintaining a particular body of expertise—when each subject has its designated "owner"—groups perform better [13, pp. 185, 223, 236–237].

Assigning roles to decision group members introduces systematised strategies of thought into decision-making that also serve to shape the ideas that come from

E. Bellos, *Sustainability Thinking*, https://doi.org/10.1007/978-3-032-11244-6_8

citizen perspectives. Donald Schön's observations about frames of thought and roles in his books *The Reflective Practitioner* and *Educating the Reflective Practitioner* hold several important insights. When an expert practitioner reflects *in* and *on* her practice, the possible objects of her reflection are as varied as the types of problem situations before her and the systems of knowing-in-practice which she brings to them. She may pay attention and reflect on the tacit norms and understandings which underlie an existing judgement, or on the theories and strategies implicit in a pattern of behaviour. She may reflect on the feeling for a situation which has led her to adopt a particular course of action, on the way in which she has framed the problem she is trying to solve, or on the role she has been assigned or constructed for herself within a larger organisational context. Reflection-in-action in these several modes is central to the art through which expert practitioners sometimes cope with the challenges of practice. When the problem situations at hand elude the ordinary categories of knowledge-in-practice, presenting themselves as unique or emergent, the practitioner may surface and question her own initial understanding of the phenomena, construct a new description of them, test them by means of on-the-spot experiments, and sometimes arrive at a new theory of the problem by articulating an intuition she has about it [233, pp. 62–63].

The mix of expert, peer, and novice has important implications in group decision-making. A disturbing finding is that groups often do not improve on and sometimes do worse than their average or median member. Due to *information signals* and *reputational pressure* four types of errors may result: (1) at times group deliberation amplifies individual errors beyond merely perpetuating them; (2) groups succumb to cascade effects as early actors suppress what could have become known from their successor peers; (3) members of deliberating groups frequently arrive at a more extreme position in line with their pre-deliberation tendencies; and (4) shared information often overwhelms unshared information ensuring that deliberating groups do not learn everything that their members know that may have been relevant. The value of role playing for the purpose of capitalising on the information that a group holds arises most prominently when all group members are informed beforehand that different members possess, and can contribute, varied and relevant information to the decision at hand. The impact of role assignment in reducing self-silencing and premature consensus can be significant in solving the challenges of eliciting distributed information. Moreover, an organisational culture that is supported by specialised watchwords can serve to surface underlying principles so as to signal that a valued perspective requires independent attention within a wider contextualisation. It licences members to provide information and viewpoints that reflect their own role—thereby ensuring that important information is taken into serious consideration. Consequently, for many decision-making groups, a crucially important task is to devise strategies that will take full advantage of the knowledge held by group members, or even by those who are not in the group [234, pp. 99, 112–113].

Four Framed Roles, Three Generic Ones

Content is specific; process is generic—Kahneman, Sibony, and Sunstein

Frame creation and selection begins with problem setting. This is the means by which we interactively *name* the things to which we will attend and *frame* the context in which we will attend to them. When a decision-participant finds himself or herself at an impasse which cannot easily be converted to a manageable problem, he or she may construct a new way of setting the problem, in other words, undertake a *frame experiment* to superimpose on the situation. And when confronted with demands that seem contradictory or inconsistent, he or she may respond by reflecting on the understandings that others have brought to the situation. Conscious of a dilemma, the reflective decision-participant may attribute it to the way in which the problem was defined, or even to the way in which his or her role was framed. In defining this role, the reflective participant may find a new way of integrating, or choosing among, the values at stake in the situation he or she is facing. These may include organisationally assigned roles and their embedded values that he or she has been asked to champion or lead. It is through the non-technical process of framing the problematic situation at issue that both the ends to be achieved and the possible means of achieving them can be clarified and organised. A professional role places basic demands on a decision-participant's behaviour, but within these constraints, each individual develops his or her own way of framing his or her role. Whether she chooses her role frame from her particular profession's paradigm and skill set, or fashions it for herself, her professional knowledge takes on the character of a system. The problem she sets, the strategies she employs, the facts she treats as relevant, and her interpersonal theories of action are bound up with her way of framing her role [233, pp. 39–41, 210].

A frame is an organisational principle or a coherent set of ideas or statements that are useful to think with. Although frames can sometimes be paraphrased by simple and elegant statements, they are actually quite sophisticated tools for thought. Proposing a frame is a social activity that can help harmonise the thoughts of the various stakeholders within a problem context. Frames include the use of certain concepts, which are assigned significance and meaning collectively. These concepts create agency to steer perceptions and explorations as a part of the creation process. Creating a frame is the result of a broader intentional action, which the frame reinterprets with a new and interesting focus. To allow reframing to occur, designers use metaphor, contextual engagement, and conjecture to "destructure" the problem and its context along with their fellow participants. The object of frames is to be attention-making, that is, they should be capable of leading to practical solutions. For frames to really inspire, they also have to be captivating. A frame should immediately conjure mental images and thoughts in participant decision-makers that trigger solution ideas through a quickfire stream of consciousness. Good frames are internally coherent and provide a stable basis for further thought, that is, a common ground for the search of solutions. The best frames are lively and thought-provoking, engaging participant's imaginations so their thoughts readily move along in the proposed direction [198, pp. 60–65].

Our proposed schema of frames has three generic roles and four specialty roles which are not mutually exclusive. The generic roles are: *self-silencing leader*; *bias mediator*; and *self-guided participant*. The four specialty roles are: *emergence thought champion, relatedness thought champion, wholeness thought champion,* and *design thought champion.* Emergence arises from the expected tension that undergirds the dynamic epistemic equilibrium that we seek to achieve between diversity and expertise (our version of the "edge of chaos") in decision-making contexts. Relatedness is fundamental to deliberation and the practice of science. Statistical correlation, experimental causality, deduction, induction, and abduction are all subinstances of relatedness. Design is the epistemic human agency needed to set the stage for practical and effective interventions whether they be public policies, programmes, projects, or institutions. It is then wholeness that brings the other three frames of thought together. Wholeness is both the outlook and due diligence required for the best application of the best available evidence, knowledge, and skills at a particular time. Indicative techniques for each frame of thought include *the design abduction model* to promote emergence; *cognitive mapping* to promote relatedness; *prototyping* to promote design; and *systems integration* to promote wholeness.

Our main focus will be on the functional role of being a thought champion for one or more frames of thought and relevant techniques. When telling/listening and demonstrating/imitating are combined and recombined, as they usually are in decision-making, they offer a great variety of possible modes of reflection that can be coordinated to fill the gaps inherent in each of these four subprocesses. Questioning, answering, advising, listening, demonstrating, observing, imitating, and criticising are chained together such that one intervention or response can trigger or build on one another [235, p. 114]. Such a chain of reciprocal actions and reflections can make up the anticipated dialogue of a thought champion and decision-participant as shown in Fig. 8.1. The thought champion must learn ways of showing and telling that fit the particular qualia of the decision-participant before her and learn how to read his particular difficulties and potentials from his efforts at thinking and performance and discover and test what he makes of her initial interventions. Conversely, the decision-participant must learn operative listening,

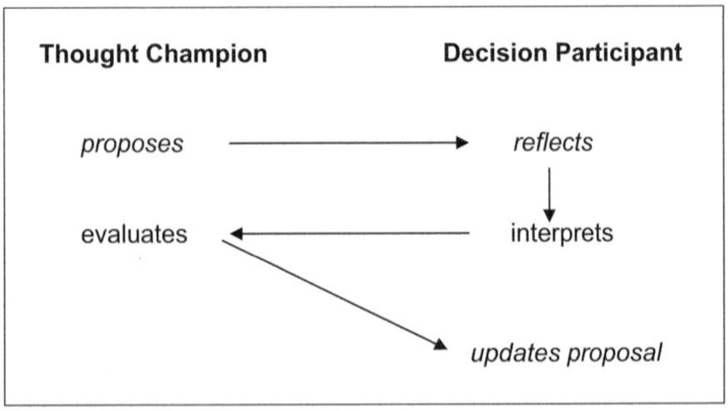

Fig. 8.1 Typical dialogue pathway between thought champions and decision-participants. (Adapted from [235, p. 114])

reflective imitation, reflection on his own knowing-in-action, and the champion's meanings [235, p. 118].

To be more effective, all seven (three generic and four specialty) roles and any of their associated techniques should be disclosed and explained to all group decision-participants. Tasks can be distributed to named individuals before, during, and the end of a meeting so that these can be tracked and assessed. Knowing the specific responsibilities and individual roles of members makes it more likely that curiosity and attention will be piqued and healthy group dynamics promoted.

Integration by the "Self-Silencing Leader" (and Team)

Rethinking is a skill set, but it is also a mindset—Adam Grant

In sustainability studies, systems integration has led to fundamental discoveries and sustainability actions that would not be possible by using conventional reductionist, compartmentalised, or disciplinary approaches. These include findings on complexity and emergent properties; interconnections among air, climate, energy, land, food, and water; assessment of multiple, often conflicting objectives; and win-win interactions, for example, that economic efficiency can be enhanced while environmental impacts are mitigated. Furthermore, systems integration facilitates clarification and reassignment of environmental responsibilities among producers, consumers, and traders; mediation of trade-offs and enhancement of synergies; reduction of conflicts; and design of harmonious conservation and development policies and practices [236].

Perhaps the most representative case of sustainability systems thinking is *Panarchy*, which may consist of both natural and socio-economic subsystems made up of a nested hierarchy of adaptive cycles. Similarly, System Implementation Reviews monitor the status of ecosystems and the effectiveness of major resource regimes. Their purpose is to provide continuous, detailed, credible feedback on the course of targeted human-environment relations so as to reveal any mismatches and produce the evidence required by would-be reformers. If deemed to be credible and unbiased by an affected community, the evidence may serve as a necessary condition for the success of reform movements [37, p. 80]. Another good example of systems thinking comes from medicine. Cochrane Systematic Reviews, now a widely accepted protocol, take a systems perspective by means of evidence synthesis. Such a review attempts to identify, evaluate, and synthesise all the available empirical evidence that meets eligibility criteria to answer a specific research question. A review specifies the question that the authors are addressing, detailing the criteria against which existing research will be assessed for inclusion. Researchers use medical conventions to define the health problem at issue and the intervention under investigation, how benefits and harms will be measured, and the type of study design that will be used. Explicit methods are used and made public with the aim of producing more reliable findings to inform decision-making. Resultant literature is catalogued and stored in web based and electronic forms made possible by the Internet [237, p. 4]. In the social sciences, Campbell Systematic Reviews attempt to emulate their Cochrane cousins but are otherwise an independent offshoot. In climate science, researchers are currently engaged in building an "evidence bank" of what climate policies work and under what conditions with the aim of informing the next IPCC synthesis report scheduled to be published in 2029. The most recent assessment was based on more than 400,000 climate change studies, many of which came from "grey" literature such as government reports. What is still needed is a systematic comparison of research approaches to gain a clear view of the evidence on which available climate policies rely, as well as designing policies tailored to an individual country's circumstances [238].

In group decision-making, studies suggest that it can be useful to appoint a *meta*-knowledge champion who can be responsible for keeping track of what others in the group know and ensuring that the group's "mental directory" stays up to date [13, pp. 236–237]. Beyond such administrative functions, additional real-time responsibilities fall on the decision group-leader. His or her role in our schema is to champion open dialogue and shepherd all thought champion and participant inputs along the themes of emergence, relatedness, design, and wholeness. Although there is no particular order or combination in which these should be surfaced and then cultivated for further deliberation and actioning, at the back of the self-silencing leader's mind should be a dynamic plan to integrate inputs so as to make a coherent, group narrative along the lines of systems thinking. Broadly speaking, systems thinking has expanded into the social realm boosted by developments in complexity theory. The origins of complexity theory lie in chemistry, chaos theory, and mathematics, but this creates a challenge to stability-based orthodoxy because it foregrounds the importance of instability, discontinuity, and non-linearity. Consequently, soft systems methodologies have been incorporated in social science toolboxes to address complicated and complex problems, such as power relations in social systems—usually an ignored dynamic in hard and soft systems theories and methodological pluralism. This has led to an expansion of methods to include participatory systems methods. As systems approaches cover various tools and methodologies from causal loop diagrams, stock and flow simulations, participatory system mapping, group model building, cognitive mapping, mediated modelling, and strategic choice approaches, among others, there is a lot to choose from both quantitative and qualitative approaches [96].

The three generic roles and four frames of thought in our schema can be interpolated in any combination into real-time group deliberations. The main task of the self-silencing leader is to assess after-the-fact which of the inputs from method specialists corroborate, elaborate, complement, or contradict each other [239], and how to integrate these with generalist participant inputs. But a precautionary note is also in order: the Procrustean expert who espouses only the methodology that he or she happens to know should be de-risked by other experts' complementary skill sets. In the end, the artistry of the thought leader, and how his or her team hold out the final decision of the group, sets the path-setting conditions for the success or failure of their group's contributions, if and when, the final decision-taking authority lies elsewhere.

The Emergence Thought Champion

I just painted the images that rose before my eyes. It is for other people to find hidden meanings in them—Pablo Picasso

The concept of *emergence* is often captured by the phrase "the whole is more than the sum of its parts". In mathematics this is known as a *superadditive* function [33, p. 233]. Evolution itself is an emergent property and so is the "invisible hand" of the market. In conditions of emergence parts adapt rather than follow fixed rules. Adaptation ensues at the level of individuals (parts) or types (wholes) [40, p. 25]. A significant challenge for decision-making emergence is that it remains hidden and becomes visible only when conditions are right. This is consistent with the views of the preeminent knowledge management expert, David Snowden, who maintains that: (1) knowledge can only be volunteered; (2) we only know what we know when we need to know it; and (3) we can always know more than we can tell, and we will always tell more than we can write down [240]. All three conditions prime emergence. A lot of ink has been spilled in the comparison of implicit versus explicit knowledge to explain, and often to counter, the proposition that "knowledge can only be volunteered". If this were not the case, the discipline of knowledge management and consultants specialising in knowledge transfer would be largely superfluous. In human beings (and some animals) hidden characteristics are expressed consciously or unconsciously all the time by an elaborate means of signalling. Consider the reality of conspicuous consumption—the man driving an expensive car, the woman wearing expensive jewellery—as a means of signalling their wealth or status. Similarly, we might infer that university degrees in general, and maths and science degrees in particular, signal a person's ability to acquire scientific knowledge. This usually translates into the higher pay that graduates earn due to the signalling value of academic degrees rather than what graduates have actually learned. The cost of obtaining an academic degree is just one example of the costs embedded in some signalling which can be made less costly by making signals more socially functional [32, pp. 303–304]. For example, choosing to consciously reveal a relevant talent, newly acquired skill, or pre-existing knowledge may lead to the emergence of better group outcomes since it would not be known a priori whether hidden talents, skills, or information have the potential to contribute positively to decision-making.

Signalling, however, can also have a deleterious effect. Informational signalling in group decision-making can lead to self-silencing when revealed preferences by the majority or by the group's leadership cause some to defer or withhold important information. If the group contains one or more people known to be experts or who otherwise have positive reputations, then the rest of the group members may silence themselves out of respect for the real or perceived subject matter expertise of these influential members. In the extreme case, sole dissenters may go as far as agreeing that a false proposition is true despite the evidence of their own senses. Self-silencing is consistent with rational self-interest when the benefits of a wise decision accrue disproportionally to the group as a whole in comparison to a self-silencing participant [234, pp. 34–35, 38]. Any resultant collective latent knowledge that emerges from group dynamics comports with the notion that "we only know what we know when we need to know it". The third and last of Snowden's points "we can always know more than we can tell, and we will always tell more than we can write down" is the most amenable for promoting strategies of emergence. But a fourth element is missing—the import of *moral* or organisational *authority*—which can be something as simple as whether one is officially empowered to take or contribute to group decision-making, or as complicated as who actually takes charge in an emergency situation.

Design Abduction: An Example of an Emergence Thinking Technique

Imagining the unknown inspires us to find new ways of existing—Joseph Le Doux

In his book, *Frame Innovation*, Kees Korst has proposed *design abduction* as a distinct category of thinking beyond traditional abduction and also beyond induction and deduction. This is the method by which working backwards from something that is only partially known or expected, a new frame can be developed to revisit the remaining unknowns of a problem situation. In design abduction the starting point is that we only know something about a hypothesised "outcome". Consequently, the challenge is to determine "what" new elements to create while not knowing the "how" of the pattern of relationships that may lead to the desired outcome. This double creative step calls for solutions for both the "what" and the "how" of the problem, and subsequently, for their joint testing. Framing is key to design abduction and involves thinking around a paradoxical situation rather than confronting it head on. Surrounding the paradox is its context, and it is here that fruitful avenues to move forward can be found. The context ultimately defines what practices will work, but new approaches can be developed to address complex, dynamic, and networked challenges without prematurely determining the problem parameters [198, p. 13]. But if we are to move from design to practice and be open to the emergence called for, is it sufficient to view problems and plan interventions as per Dorst's "double creative step"? The nature of emergence is such that an outcome can only be imagined, which implies that in complex contexts none of the parts in Dorst's three-part equation (What + How = Outcome) would be explicitly known ahead of time. We can in fact expand non-linear thinking by adding a longer ascending "results chain". This is comprised of inputs,

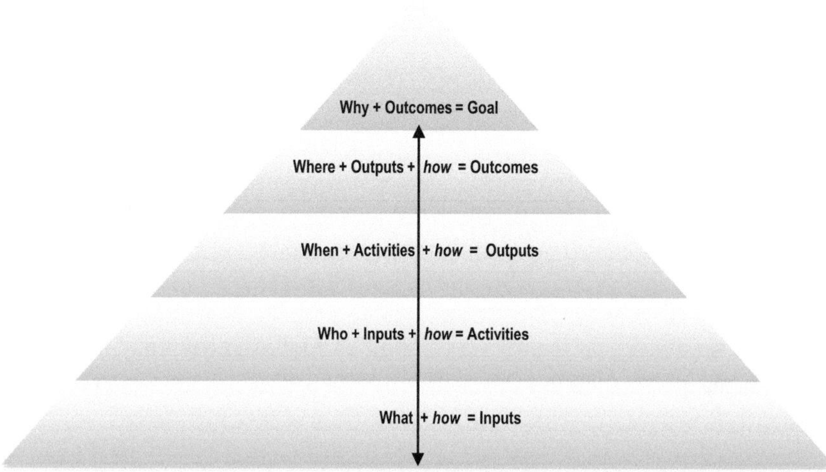

Fig. 8.2 A design abduction framework for decision emergence

activities, and outputs before approaching a desired outcome as shown in Fig. 8.2. Here the ascending "how's" stand-in for the "programme theory" concept that is popular in the discipline of monitoring and evaluation. This ladder of *how's* can be used to create open-ended dialogue in the spirit of design abduction. Design thought champions may consider what resources would be needed to successfully complete an intervention; who would lead and manage the intervention; when could the intervention be realistically completed; where the reach of the intervention would end; and why the intervention is significant. When the thought champion engages in real-time reflection in situations he or she perceives as unique, directing attention to phenomena of interest and thereby surfacing an intuitive understanding of them, his or her experimentation is at once exploratory, action testing, and hypothesis testing all in one [235, p. 72].

When decision-making is opened to emergence, the final outcome from such processes may actually be counter to what was originally envisioned. Then, paradoxically, the solution is even more emergence as implied by the bi-directional arrow in Fig. 8.2. In other words, the model can be "run" again. Even if de-limited by time or scale, emergence can still yield numerous ideas. Thus, the *first* of intuition becomes essential to evaluate any uncomfortable results as to whether initial options are acceptable or not. Viable ideas can then both inform and rely on the more rigorous decision-taking that comes from wholeness. Importantly, design processes that include implementation also create a feedback loop of information, ideas, people, action, and reaction [96, p. 48]. It follows that iterative feedback to promote learning and self-correction can be designed-in so as to be context adaptive and risk responsive, in other words, to be emergence sensitive.

The Relatedness Thought Champion

Genes can influence organisms at a distance—Nicholas Christakis

The modern world has depended on horizontal societies of mind to guide and grow a knowledge commons. The Royal Society was founded in London in 1660, and by 1880, there were 118 learned societies devoted to science and technology with nearly 50,000 members—part of a rich variety of organisations—some arranged around general interest, others around specialist subjects or hobbies, some concerned with professional certifications, and others about advancing the frontiers of knowledge. The now familiar use of academic journals, conferences, and mutual commentary was used to improve the design of new ships, improve the state of medicine, and advance physics [24, p. 56]. Scientific analysis by correlation or causation is a specialist demonstration of relatedness. The resultant web of knowledge is variously configured by *degrees of strength* among its constituent parts, a factor which aligns in the most basic way with nature. Genetic and developmental processes determine the connections among neurons, but they do not specify the strength of those connections. Strength—the long-term efficiency of synaptic connections—is regulated by experience and learning. Learning selects from a large set of pre-existing connections and alters the strength of those connections [174, p. 202]. In the proposed schema *connectedness* exhibits the strongest degree of strength at the bottom of a three-level system, but less so moving up to *nearness* with its associations of intermediate strength, and finally up to *relatedness*—with the weakest strength but also the highest number of associations encompassing all the rest. This is similar but also different from "Dunbar's number". In the 1990s anthropologist Robin Dunbar proposed that the larger hominid brain evolved to be able to cognitively maintain up to 150 social interactions with decreasing strength the further one goes away from a cohesive 5-member family unit [241]. The relatedness strategy proposed here is also made up of variable group sizes and interaction strengths but is more expansive because it is based on, but not limited to cognitive associations—it includes *informational associations*. And unlike the Dunbar number and Metcalfe's Law, it also involves *institutional associations*. Multiplied by all potential participants, a supraplane of relatedness associations may be established that is led by self-organised institutions. An institution is not merely a brick-and-mortar structure housing, say, a bureaucracy like a post office or a union hall, it is a compilation of working rules that are used to determine who is entitled to make decisions in some arena, what activities will be allowed and what activities will be constrained, what aggregation rules will be in force, what procedures are required, what information is or is not required, and what payoffs will be available to individuals dependent on their behaviour. Working rules are common knowledge and are monitored and enforced, but need not be the same as enacted laws, administrative regulations, and court decisions [126, p. 51]. Finally, the proposed schema is dynamic instead of being static allowing for a mobility that creates *new associations*.

Neighbourhoods are examples of nearness, while other groupings, say, families represent connectedness. Relatedness differs from nearness in that it is able to address "spooky action at a distance" to borrow Albert Einstein's description of quantum entanglement. It is also different from connectedness because not all of its components need to be actually coupled, and multiple dimensions are possible in contrast to traditional network models. We can therefore say that relatedness differs both horizontally and vertically from connectedness. Relatedness has also in part been interpreted as spatial heterogeneity which is information bearing since it reveals both the intensity and pattern of spatial associations. Spatial analysis and geographic information science continue to develop sophisticated methods for extracting explanatory and predictive power from geo-spaces [242]. Spatial analysis may also have special significance in particular places. In South Africa, for example, the legacy of apartheid on the country's spatial patterns continues to be felt to this day. Geographers have creatively applied Tobler's First Law of Geography which states: *everything is related to everything else, but near things are more related than distant things*, and in so doing, have applied the idea of nearness as a refinement of "everything is related to everything else". The conception of nearness is supported by complexity theory which highlights the importance of geographic context in that a system's growth and development is sensitive to the intensity and pattern of local (near) interactions. Nearness has also been interpreted as the "shortest-path relations" among all pairings. These are the minimum-cost routes for physical movement or virtual interactions between objects, when cost is assigned generally. The shortest-path relations can be used to refine the measurement and analysis of geographic and other attributes [32, p. 121], [242]. Based on Tobler's law, we can further propose that nearness can be measured by distance when it is defined as distance from origins (ontology), or distance from pre-determined objectives (targets). Distance from adopted public policy targets has important implications for policy analysis, programme management, and monitoring and evaluation. Defining and measuring the "distance" left to achieve pre-determined objectives can not only improve oversight, but also the comparison of relative progress among similar programmes and projects carried out contemporaneously. In the realm of sustainability, the challenge of overall fit concerns the compatibility between the management of ecosystems and the institutional arrangements established to oversee human activities affecting those systems. Generally, there is a presumption that the closer the fit between ecosystems and institutional systems, the better the relevant institutions will perform [37, p. 20]. In short, the relatedness supraplane is constantly changing to reflect dynamic informational associations within more stable governing institutions.

Connectedness—the strongest association in our schema—is well represented by network science. Economic relations, the Internet, ecological webs, technological networks, information networks, transportation networks, and brain connectivity are some of the areas of interest that have been represented and modelled as networks of interconnected parts [243]. Social networks are a favourite focus of research and offer a rich variety of subjects ranging from hunter-gather societies to online gaming communities [28]. A network consists of *nodes* and *edges* that link them. Nodes that are joined by an edge are referred to as *neighbours* and are considered to be *connected* if it is possible to get from any one node to any other along edges. Edges can be *directed* on *undirected.* In an undirected network, the *degree of a node* equates to the number of edges that connect to it. The *average degree* and the *degree distribution* tell us, respectively, the average number of node connections, and to what extent a node is interconnected as compared to others. *Path length* refers to the minimum number of edges that must be crossed to get from one node to another. Nodes on minimal paths play critical roles. A node's *betweenness* is the percentage of minimal paths that go through it. A final statistic, the *clustering coefficient* is the percentage of a node's pairs of neighbours that are also connected by an edge. Networks can be represented as lists, matrices, or graphs. Graphical representations of networks are the most popular, but matrices and lists are more suited to calculating network statistics [32, p. 118].

Network structures include Small World, Power Law, Random, Geographic, and Hub and Spoke shaped networks [32, p. 122]. A network's structure together with its dynamic functions constitutes its *topology.* A promising new concept known as a "network morphospace" is a computational and visualisation method of network architectures that helps to identify regions enclosed in this space that contain existing networks as opposed to empty regions in which network designs do not exist but may emerge as being either possible or impossible options. It is feasible to add fitness as a dimension of the morphospace thereby converting it into a fitness landscape [243]. We can analogise further by reconsidering empty regions of the morphospace as being similar to *basins* in the landscape model, or what Ron Burt calls *structural holes* in network theory [32, p. 129]. People who occupy central positions in networks are able to fill structural holes and may in consequence be considered to be influencers or decision-makers in the network. The locus of final decisions, however, may lie elsewhere. Widely distributed networks are good for deliberation and argumentation as well as collecting information. They grow and maintain a knowledge commons and communities of practice, but they work less well in converting decisions into action. These functions tend to be dominated by small groups, with a strong common subject matter understanding and culture. They combine wide networks with much smaller core groups of gatekeepers, guardians, curators, and editors who shape contentious issues and arrive at final decisions [24, pp. 57–58].

Cognitive Maps: An Example of a Relatedness Thinking Technique

The very fact of interrogating the concept of reason using reason presupposes the validity of reason—Steven Pinker

A cognitive map is a semi-quantitative model of how a system works. Cognitive maps are especially applicable and useful tools for modelling complex relationships among several, and sometimes disparate, variables. The map is based on key identified variables and the cause-and-effect relationships among these variables. These can be physical quantities that can be measured, such as percent of vegetation cover, or complex aggregate and abstract ideas, such as political factors and community values. The lead person making the cognitive map decides what the important variables are that affect a system and then draws causal relationships among these variables showing the relative strength of the relationships with a number between -1 and 1, making them “fuzzy cognitive maps”. The various directions of the causal relationships are indicated with arrowheads and may include feedback relationships. Fuzzy Cognitive Maps are relatively quicker and easier to acquire from human sources, who do not usually think in equations. Many knowledge sources with diverse knowledge and different degrees of expertise can be used. There is no restriction on the number of experts or on the number of concepts. With cognitive mapping one decision-maker’s map can be analysed and compared to another’s and the most useful one selected, or the maps can be combined. Maps can be made with local and indigenous people, who often have quite a detailed understanding of, say, their particular ecosystem. Their inputs can be valuable for decision-making, as well as for a community to accept the chosen solutions. In addition, the effects of different policy options can be easily modelled. To analyse a cognitive map, the number of variables and the number of connections in the map can be counted. Cognitive maps are directed graphs, or digraphs, as derived from graph theory. According to graph theory, cognitive maps can be converted into adjacency matrices [244] and several indices, such as centrality can be calculated. This measure shows how connected the variable is to other variables and what the cumulative strength of these connections are. Graph theory provides us with many more indices in addition to the number of variables and connections [244] (See Fig. 8.3).

The representation of temporal and spatial scales is the weakest aspect of fuzzy cognitive maps but can be improved by linking them to geospatial models, either directly or through the use of scenarios. Planning scenarios, and especially future narratives of normative strategies, often remain qualitative and based on current systems understanding. Consequently, methods to construct fuzzy cognitive maps of perceptions of future systems could play a role in today’s discussions of societal transformation to meet the original goal of limiting global warming to 1.5 °C [245]. As a semi-quantitative technique, fuzzy cognitive maps find a place between causal loop diagramming and system dynamics modelling [239].

The Design Thought Champion

System visibility is a necessary precondition and enabler for sustainability design—The Karlskrona Manifesto

Design is the process of understanding the world and articulating an alternative conception on how it should be shaped according to the designer’s authority and intentions. By design, we cause change and shape our environment. If we don’t take

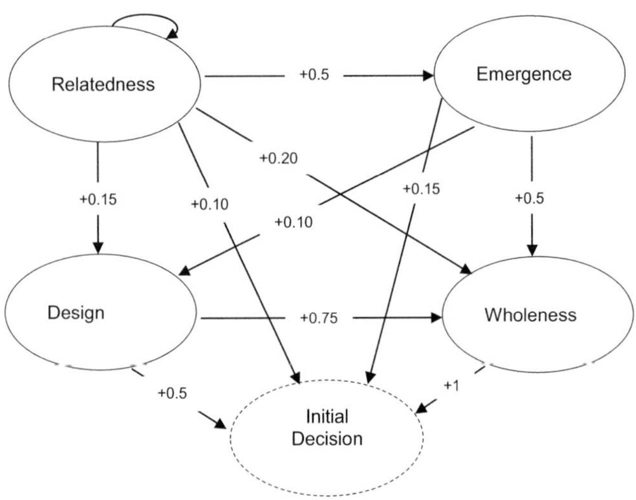

Adjacency Matrix of the Four Frames of Thought

	Relatedness	Emergence	Design	Wholeness	Initial Decision
Relatedness Strategy	0	+ 0.50	+ 0.15	+ 0.20	+0.10
Emergence Strategy	+ 0.5	0	+ 0.10	+ 0.5	+0.15
Design Strategy	+ 0.15	+0.10	0	+ 0.75	+0.5
Wholeness Strategy	+ 0.20	+ 0.5	+ 0.75	0	+1
Initial Decision	+0.10	+0.15	+0.5	+1	0

Fig. 8.3 Fuzzy cognitive map of four frames of thought. (Source: Author)

sustainability into account when designing, no matter in which domain and for what purpose, the opportunity to cause positive change will be missed [246]. The practice of design has traditionally addressed complexity by problem reframing, working with inconclusive outcomes, incorporating cultural practices, and across disciplinary boundaries [96]. In sustainability, human and natural systems are coupled and evolve over time as complex adaptive systems with emergence, evolution, and adaptation varying by scale. Accordingly, the development and quantification of frameworks along organisational, spatial, and temporal dimensions is needed to avoid sustainability solutions in one system causing deleterious effects in other systems [236]. This process can begin with design diagnostics. Table 8.1 sets out several design implications for environmental institutions revolving around ecosystem properties, actor attributes, and implementation issues. These simple design diagnostics can be analysed mechanically, but the expert diagnostician and talented design thought champion can bring an intuitive feel and understanding to the situation and act accordingly to resolve each specific problematic condition.

Table 8.1 Design implications for environmental governance

Diagnosed condition	Design implication
Non-linear or chaotic systems; surprises	Early warning devices/rapid response capability (e.g., real-time monitoring)
Problem duration	Management structures (e.g., dispute resolution structures and procedures)
Functional interplay	Coordination mechanisms (e.g., institutional links across scales)
Uncertainty; imperfect knowledge	Social learning; adaptability; precautionary approach (e.g., ongoing risk assessments)
Variability of political and socio-economic systems	Flexibility (e.g., latitude of individual members to implement decisions)
Heterogeneity of member interests	Issue linkages (e.g., national and international "package deals")
Asymmetries in causal responsibility	Emphasis on equity (e.g., global north versus global south disparities)
Asymmetries in capacity	Capacity building; technology transfers (e.g., climate mitigation and adaptation funding)
Violation tolerance	Systems of implementation review (e.g., rapid evaluations under low tolerance conditions)
Incentives to cheat	Sanctions; deterrence (e.g., legal accountability)
Lack of transparency	Monitoring procedures (e.g., independent monitoring and evaluation)
Malleability of rules	Certification; authoritative interpretations (e.g., reputational determinations and records)

Source: Adapted from [37, p. 178]

In decision-making groups, design thought champions can help put things together and bring new things into being, and in the process, confront emerging variables and several constraints, some initially known and some discovered through the designing process itself. Designers have to prioritise variables, reconcile conflicting values, and navigate around constraints—a process in which there are no unique right answers. The ensemble of problem framing, on-the-spot experiment, detection of consequences and implications, feedback, and the response to it, which constitutes a reflective conversation with the conditions of a situation, creates the design-like artistry of expert practice and its decision-making [235, pp. 42, 158].

Prototyping: An Example of a Design Thinking Technique

> *The rate of the development of science is not the rate at which you make observations alone but, much more important, the rate at which you create new things to test*—Richard Feynman

Design thinking is a pre-factual process that is useful in *prototyping*. Prototyping of public services is one means of testing emerging ideas to help choose between alternatives. It is a flexible methodology that can be applied to new services or improve existing ones. The value of undertaking prototyping is typically worth the extra cost, as it ultimately reduces the risk of catastrophic failure. Prototyping also serves to better understand the nature of the problem itself and to build confidence among stakeholders that impactful solutions are being developed [96]. Prototyping can be thought of as the first draft of a document—it sets the stage but the exact contours of the final draft cannot be accurately foreseen. The design thought champion can help set out some prototyping options to help create appropriate interventions. For example, patterns of modern energy use are significant drivers of climate change; therefore, a prototyping matrix for possible energy services policies and programmes may reveal areas for attention that were previously missed. To this end, Fig. 8.4 features several considerations for further design actioning.

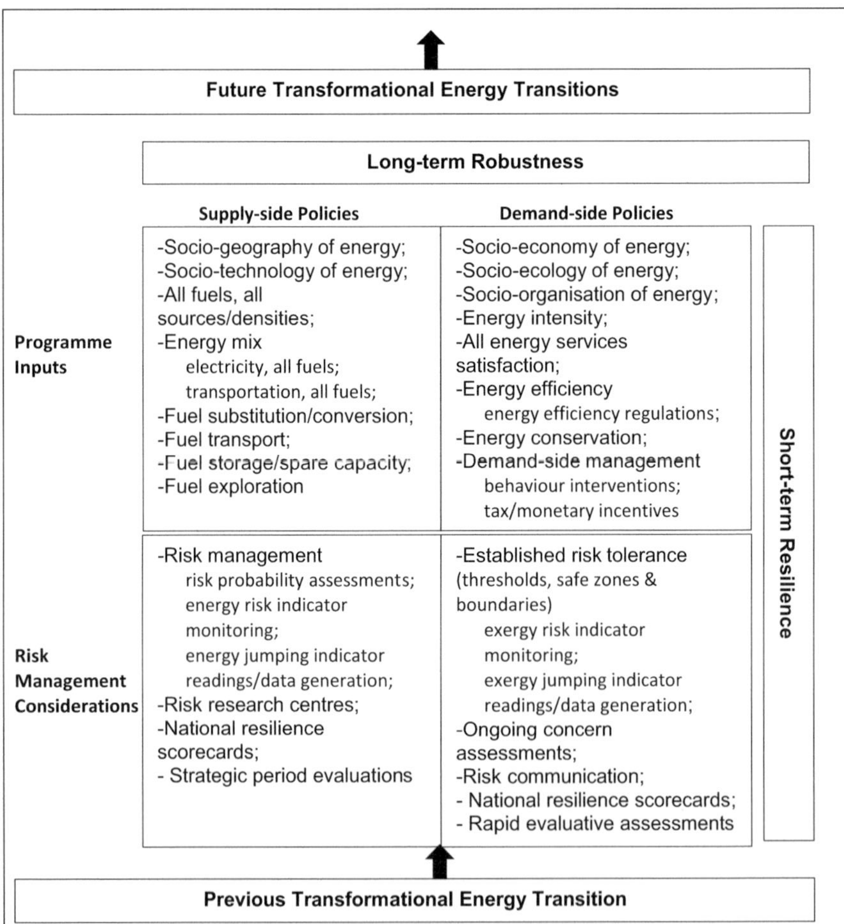

Fig. 8.4 Concept-to-design prototyping matrix for energy services. (Source: Adapted from [103])

Having identified a menu of several *socio-systems* and related considerations, it would then become necessary for decision-makers to prioritise and select the specific items that are the most relevant for their particular contexts. Ideally, each item chosen from the top left cell of Fig. 8.4 should be considered together with other suitable items from within the cell and then be combined with as many feasible options from each of the remaining cells at the most practicable granular level. For example, a cross-tabulation of "socio-technology of energy" with that of "energy efficiency" has enormous scope to improve conversion efficiencies, particularly for appliances and buildings. And a low demand-side energy policy suggests how an energy service lens opens up new opportunities for progressive action—effects that are not commonly resolved in global scenario and modelling analysis [247]. Similarly, resilience and robustness metrics become more useful when individually calibrated and compared against selected standards. Ultimately, the proposed matrix itself can be modified or enhanced as long as the principle of considering multiple cross-impacts is maintained. The final results of these exercises cannot be foreseen in detail hence the value of thinking ahead via prototyping, including paying attention to any possible emergent properties that can then be tracked for subsequent adjustments.

The Wholeness Thought Champion

Life in all of its evolutionary wisdom, manages ecosystems of unfathomable beauty, ever evolving towards wholeness, complexity and consciousness—Frederick Laloux

The idea of "wholeness" is not a new concept in science or philosophy and has found expression in different forms, such as the integrative functions of holons and sub-holons we encountered earlier. Probably the most familiar yet powerful illustration of wholeness is the human capability to recognise faces. This is accomplished by being able to see the configurational aspects of the whole as opposed to its parts. Faces are broadly similar complex three-dimensional structures that have to be differentiated from one another by miniscule interrelations among the parts and individually discriminated in different orientations, at varying distances, in different lighting, and often while moving quickly, yet we are able to do so. The inability to recognise faces is a medical condition called *prosopagnosia* [5, p. 60]. Christopher Alexander has gone so far as to experimentally demonstrate that wholeness is not actually visible to architectural students [248, pp. 453–457]. But the late David Bohm has perhaps been the most eloquent and consistent advocate of wholeness. He has suggested that the parts of a whole can be physical objects or mental events, including creative acts of intelligent perception. In his view, inanimate matter, living organisms, and "mind", all arise in a single space in which they are enfolded, or contained implicitly [249]. This is what is meant by "implicate order" in his book *Wholeness and the Implicate Order*. Even discrete atoms, as constituent parts, can under certain conditions be fashioned into a whole. Consider the discovery of the

Bose-Einstein condensate by scientists in 1995—a fundamental form of matter over and above the previously recognised gases, liquids, and solids—in which many atoms act as a single atom in one quantum state [23, p. 4]. The various sciences occupy different hierarchical levels which together form a single interconnected structure. The coherence of this structure arises out of the associations among its parts. One way to obtain greater coherence is by identifying a "universality class" which denotes that one phenomenon does not depend on the details of the concept crafted to explain it, but rather may draw from several incomplete or even wrong models to explain how the real-world works, provided the real world and the models lie in the same universality class [250, p. 284]. We can think of "partness" and "wholeness" as two basic universality classes. Partness is the dominant approach in science—it manifests itself in *reductionism*. Conversely, *wholeness* involves the understanding and application of cross-cutting laws of science such as the laws of thermodynamics.

The integration that is needed to establish wholeness in social systems begins with our biology. The fundamental activity of the nerve cell is indeed integration: it sums up inhibitory and excitatory synaptic potentials as they are received from pre-synaptic neurons and produces an action potential only when the total number of excitatory signals is greater than a minimum threshold of the total number of inhibitory signals [174, p. 93]. Integration in human social affairs begins at an even deeper level—our genes. The effect of genes within a particular individual often depends on the actions of other genes in a process called *epistasis*. Genes get expressed at multiple levels from protein synthesis to physiology to behaviour. Imagine your genes predisposing you to play matchmaker such that a friend with attention-grabbing blue-eyes (which evolution must have favoured over brown eyes in certain instances) prompts you to introduce her to another one of your single friends, then this social web that has been created ultimately affects your own survival in a process that has been called *social epistasis*. Thus, we can expand the "within-body genetics" that has been limited to body and mind in evolutionary thinking and take up new ways of seeing human social systems [28, pp. 335–349].

We navigate social complexity guided by our social brains. Importantly, our mental circuitry for decision-making is integral with our circuitry for memory. It has been found that short-term memory is derived from operational changes in the synapses of a neural circuit, whereas long-term memory is based on structural changes [174, p. 221]. And whereas short-term memory has limited capacity, it is dynamic, flexible, and is readily accessible. Long-term memory, however, is stable without requiring active attention and has vast storage capacity for new information [164, pp. 12–15]. Memory acquisition can be summarised as a dynamic process leading to a stable structure just as human language displays structural elements (Broca's area of the brain), while the process of language acquisition in human infants, for example, has been conserved by evolution with its dynamic nature often on display at the conscious level in the effortless way it can produce spontaneous novel intuitions and inventive speech at the speed of thought [2]. Daniel Kahneman in his book *Thinking, Fast and Slow* popularised the dual-process System 1 versus

System 2 dichotomous model of human cognition. System 1 is fast, intuitive, relying on shortcuts such as "anchoring", "representativeness", and "salience". This can be contrasted against System 2 capabilities which are slow, logical, and statistically literate [85, pp. 21,24]. System 2 is often described as providing the necessary overriding executive functions from its location in the brain's neocortex. But this conception has been refined by a more holistic metaphor such that scientists now talk of integrated perception/memory streams. Humans have the ability to use working memory to integrate inputs across sensory modalities, to think schematically and conceptually rather than just perceptually, and to orient towards the future. We experience our lives as unified events and streams rather than a mere collection of sights, sounds, and other isolated perceptions [25]. Moreover, these processes can be reiterative according to Martin Seligman and his collaborators who say

> …there aren't two independent processors. Rather, deliberation depends fundamentally on intuitive affective evaluations. The same intuitive processes that deliver affective evaluations of ongoing actual situations are *reused* during deliberation, but they are now directed at situations constructed in the mind. We call this view, which understands intuition and deliberation as deeply interconnected, the "massive reuse model" [251].

Simon French and Nikolaos Argyris mirror and collectivise Daniel Kahneman's systems distinction such that *System 1 Societal Deliberation* consists of informal discussions with no formal governance between decision-makers, experts, stakeholders, and others concerning a decision and *System 2 Societal Deliberation* which involves formal deliberations and decision-making set within defined governance structures and constitutions that control who may take part, their responsibilities and authorities [252]. It follows that the next developmental step is to reframe these fragmented analytic parts into what we might call the *massive societal reuse model* as it relates to decision-making, which is analogous to *process* that can be conjoined to its *outcome*—a major departure from the two existing frameworks that calls for the unity of each separately and jointly.

Systems Integration: An Example of a Wholeness Thinking Technique

> *It is not among extraordinary and fantastic things that excellence is to be found*—Blaise Pascal

Causal loop diagrams are the most frequently utilised method to visualise systems by describing existing causal relationships among selected sets of variables together with their positive and negative feedback loops. Positive feedbacks amplify a causally related variable in the same direction, while a negative feedback dampens a causal relationship in the opposite direction. The UK Cross-Government Trial Advice Panel, which supports experimental design to understand whether programmes and policies are effective, emphasises the need to understand and manage

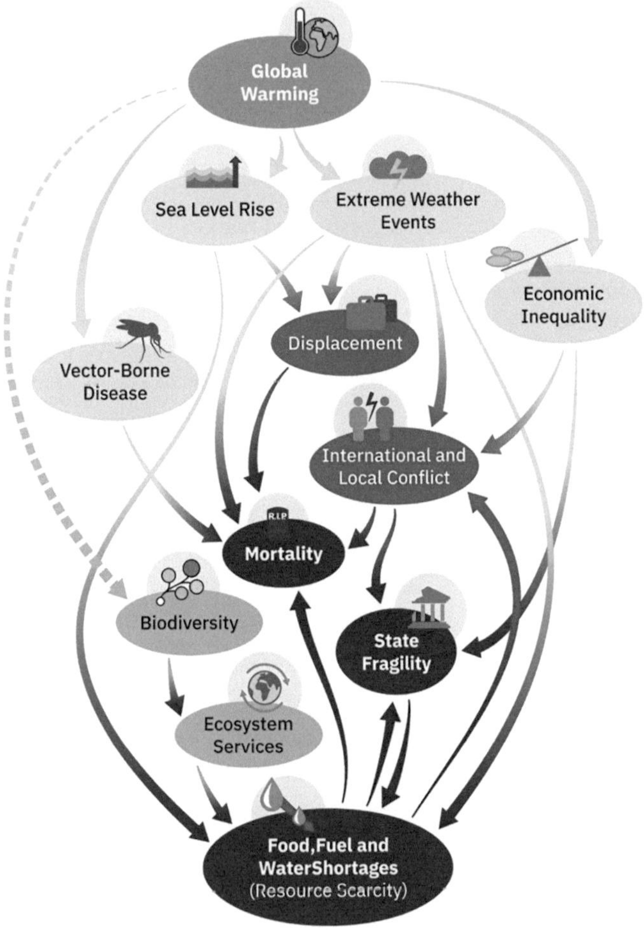

Fig. 8.5 Causal loop diagram of risk cascades from global warming (solid lines denote amplifying feedback, dotted lines denote dampening feedback). (Adapted from [253])

feedback loops[1] [96]. Figure 8.5 is a "user-friendly" sample of a causal loop diagram related to global warming. Causal loop diagrams are often the first step in visualising complexity and can set the stage for more predictive methods like dynamic systems modelling.

System dynamics models can build on causal loop diagrams to quantify and simulate a wide variety of phenomena ranging from an economy to an ecology. Such a model would consist of *sources, sinks, stocks, flows, rates,* and *constants*, all

[1] Useful resources include Donella Meadows' *Thinking in Systems and* the OECD's report titled *Systems Approaches to Public Sector Challenges.*

with their specific symbols.[2] Sources provide inputs into system stocks; stocks hold a level of variables; sinks absorb outputs from the flow of stocks; and flows detect positive (amplifying) or negative (dampening) feedbacks between levels of stocks. The level of a stock changes based on sources and flows. Rates and constants address these flows, which can change over time, while sinks and stocks serve as placeholders for any processes external to the system being modelled. Even if the direct effect of increasing or decreasing a rate of flow increases or decreases a stock, positive and negative feedbacks in the system imply that other stocks will also change values, so the net effect of a change in a rate or flow may be reduced, cancelled, or even possibly reversed [32, pp. 201–212]. Stocks and flows and their feedbacks have wide application. Without such a framework the concept of carbon trading and carbon accounting would not be feasible. Similarly, Greenhouse Gas Inventories are emission inventories (stocks) of natural and anthropogenic greenhouse gases from countries, regions, and cities that scientists use when developing atmospheric models. They typically combine the various greenhouse gases into a single weighted value of emissions (constants). Policymakers and regulatory agencies then use these inventories to establish compliance records with allowable flows of quantified emissions (rates). In addition, stakeholders use them to better understand trends (sources and feedbacks), including removals by carbon sequestration (sinks). All Annex I countries are required to report annual emissions and sinks of greenhouse gases under the United Nations Framework Convention on Climate Change [254].

In summary, the four proposed expert roles by way thought champions concerns the requisite yet varied focus needed for problem resolution. The proposed four frames of thought can be implemented by subject matter experts who can bring their own relevant methodological skill sets *to* decision-making *for* decision-taking. Dynamism from positive and negative feedback loops in systems of knowledge translates into multitudinous words, numbers, images, ideas, and actions together with all their interactions. Starting from their formative stages, these may be variously countered by unintended consequences, faulty assumptions, inflexible perspectives, human biases and fallacies, and stressors and shocks—in other words—complexity. The proposed four frames of thought can help reduce *and* absorb complexity. *Design-wholeness* is convergent and reduces complexity by framing and scoping, while *emergence-relatedness* is divergent and absorbs complexity by recombination and conflation. The leader of a decision grouping has the final important role of integrating inputs for maximum impact. Collectivised overarching roles in our proposed schema translate into having well informed and active citizens, as well as innovative and ethical experts. At base, the shared intentionality hypothesis can be made operational in organisations and polities by means of the proffered Collective/Democratic decision model discussed in Chap. 10.

[2] Respectively: clouds for sinks/sources; boxes for stocks; arrows with plus or minus signs for flows; joined tip-to-tip triangles for variable flows; and circles bisected by arrows for constant flows.

Chapter 9
Planning Scenarios as Collective Perspectives

The most dangerous worldview is the worldview of those who have not viewed the world—
Alexander von Humboldt

The human predilection to look to the future is made practical in many ways, including by formal forecasting. Critically, the longer the time horizon, the less accurate a forecast will be. Forecasting encompasses a wide range of methods, some machine driven, such as computer simulations, some driven by traditional human intelligence. From a large and diverse pool of such sources and specimens, arguably the best example is weather forecasting (if by "best" we mean the capacity to convert complexity to usefulness). Since 1792, *The Farmers' Almanac* had offered weather predictions for the growing season, but those predictions had never been much better than guessing. In 2016, the U.S. National Weather Service's five-day weather forecast was as accurate as the one-day-out forecast had been in 2005. Modern weather satellite technology, weather radar, weather balloons, and weather buoys made a big difference. Before 1992, meteorologists had simply inputted data (wind speeds, ocean temperatures, and so on) into their computer forecasting model. However, these were just estimates as there were no planetary hard data to account for the "butterfly effect" at play. Then the human element came with the adoption of "ensemble forecasting". It called for every weather forecast to have a *cone of uncertainty*—not just for hurricanes. If one forecast in the ensemble said one thing which was contradicted by another due to trial selections of input variables, then the probability accuracies had to be adjusted according to a new statistic: how predictable the weather was at any given moment [255, pp. 150–152, 167, 181]. Of course, the "mother of all weather forecasts" concerns anthropogenic climate change anchored on the 2015 Paris Agreement which called for keeping global warming to no more than 2 °C above pre-industrial levels. However, by 2030 projected levels of greenhouse gas emissions have been estimated to grow to be more than 15% higher than the 2015 level at 55 gigatonnes and even higher than the *logistic* curve forecast for 2050. This and other growth trends have led Vaclav Smil to conclude that most economists are either dismissive or unaware of the advances made in our

E. Bellos, *Sustainability Thinking*, https://doi.org/10.1007/978-3-032-11244-6_9

understanding of the synergistic functioning of today's civilisation and the biosphere yet overreliance is placed on them to supply the physically impossible narratives of continuing growth that guide decisions made by companies and national governments [59, p. 507].

Planning scenarios may be just the right vehicle for generating *both* a participatory policy narrative and setting the stage for expert inputs like computational modelling. This comports with de Vries's view of combining storytelling and modelling. The storytelling part consists of carefully constructing narratives built around interpretations of past and current trends and observations. The modelling part is an attempt to bring about consistency and focus by quantifying certain parts of the narratives on the basis of available data and formalised relationships [256, p. 383]. Unlike projections and forecasts, which tend to be more quantitative and more circumscribed in their key assumptions, scenarios are cogent narratives dealing with possibly far-reaching changes. A scenario may include a history of developments, a possible course of events leading to an actual outcome, an imagined future, or some combination of these. Scenario analysis offers a uniquely valuable means to craft a common framework for diverse stakeholders to map and address their most pressing concerns and to identify alternative solutions, as well as to serve as a forum for discussion and debate [257]. Scenarios are typically used in the context of planning over long time horizons or short-term decisions that have long-term consequences by widening perspectives and highlighting key issues that may otherwise be missed. One of the great benefits of scenario planning is the articulation of a common envisioned future that enables more coordinated decision-making and action. Scenario development commonly targets issues which are sensitive to stakeholders and provides the means by which decision-makers can anticipate coming change so as to prepare for it in a timely and responsive manner [258]. Practitioners like Hannah Lübker have noted that there is a critical need to use one's imagination and draw on creative and dynamic stories about the future for improved decision-making in the present. Collectively held visions of the future can help to define goals and create a sense of what is possible and desirable [259]. In many parts of the world, a more participatory style of decision-making is becoming widespread, superseding authoritarian and technocratic approaches. Additional criteria, such as the environment, human rights, animal rights, and gender are now being considered in decision-making. The emergence of new social actors such as non-governmental organisations, multilateral organisations, and transnational companies leads to an increase in the number of dimensions used to define issues, problems, and solutions and therefore to higher complexity [47, p. 364], including moral and ethical questions.

The two main branches of scenario development consist of exploratory or anticipatory approaches. The former examine the future according to known processes and extrapolations from the past. The latter are more subjective and follow different or feared visions of the future that may be either achievable or avoidable. This camp includes policy-responsive scenarios in which policy decisions based on critical issues are considered with the desired policy in mind. As such, this type of scenario is more often found in governmental and organisational decision-making in the attempt to better understand and manage risks. Policy responsive scenarios may traditionally consist of either expert judgement or stakeholder-defined scenarios. Expert judgement-driven scenarios model future conditions by means of scientific knowledge derived from decisions, criteria, objectives, and rules established by scientific evidence and subject matter experts. Advantages of this type of scenario include the integration of current thinking, the incorporation of a wide range of pertinent information, and the ability to build a scientifically based consensus. Major disadvantages of scenarios dependent on expert judgement are the introduction of bias and a potential lack of political plausibility. By way of comparison, stakeholder-defined scenarios involve actual stakeholders that identify the assumptions that underlie the emerging scenarios. They usually have greater political plausibility and public acceptance than expert-driven scenarios due to the active involvement of stakeholders in the scenario development process but also suffer from the potential biases that the most active citizens bring [258]. In the realm of sustainability, scenario and visioning methodologies enable a wide range of actors to jointly explore diverse development trajectories and alterative futures based on mutually agreed upon criteria and systematic stakeholder participation. Critically, such approaches make visible the normative, ethical, and political issues that are often obscured by the purely descriptive analysis of environmental and human systems. Visions are intended to inspire and provide direction, while scenarios set out the major implications, uncertain though they might be, of different pathways into the future. Scientifically crafted visions and scenarios, however, often provide compartmentalised expert advice, focus on future constraints, and define uncertainty in hard-to-understand probabilistic terms. They fail to offer broad appealing images and storylines that also account for choice and opportunity and a richer conceptualisation of possibilities and uncertainties [156].

Based on Pierre Wack's pioneering work in the field of futures research, Sunter and Ilbury recommend a three-step process to craft a scenario: (1) identify factors that almost certainly will apply in the future; (2) identify two key opposing uncertainties and use them to construct a matrix with four quadrants; and (3) use the selected factors and uncertainties to create four different scenarios, which if creatively named, spark the imagination and guide discussions. The result should look something like Fig. 9.1. This is a visual representation of possible scenarios: a best-case in which the outcomes of both uncertainties are positive, and a worst-case scenario in which both uncertainties are negative. The other two scenarios make up the middle ground, where each positive outcome in one uncertainty meets a negative one in another [260, pp. 4–5, 63–69]. Opposing uncertainties do not necessarily imply definite causal relationships—they are defined in terms of their relevance. The different elements are then put together in the form of a narrative, showing how the world could change in the future [47, p. 366]. Climate change will be a certain driver long into the future, as is worldwide economic growth based on resources extraction. The rate of change, however, is unknown and creates uncertainties with respect to sustainability up to and including societal collapse. Most likely, synthetic biology, information technology, and artificial intelligence will drastically change economies, culture, and civilisation, facilitating new unknown futures. Technological change alone will not lead to the systemic transformations needed for global

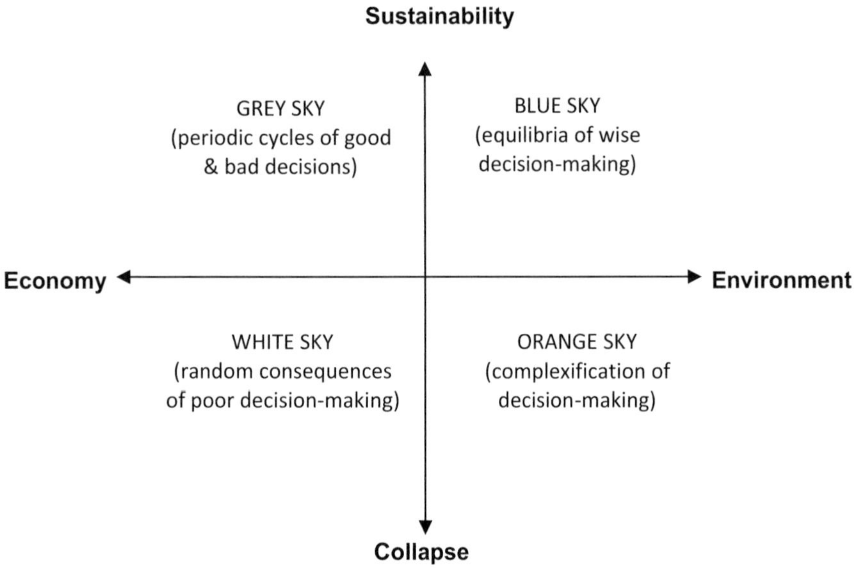

Fig. 9.1 Four global sustainability scenarios and their decision states

sustainability. It could lead humanity in different directions, both pleasant and unpleasant, and with varied social and environmental impacts [99].

In accordance with our guiding schema, the cross-tabulations of the main uncertainty dimensions and their associated decision states are shown in Fig. 9.1. Although there are several substantial drivers behind the identified uncertainties in all four scenarios, focused attention on climate change, inequality, technology, and decision-making can serve as the primary drivers for focused analysis.

- The *grey sky scenario* evokes a cloudy future. Multiple crises reinforce each other endangering the 2030 Agenda for Sustainable Development. Unsustainable patterns of growth and consumption continue to contribute to climate change and pollution with concomitant environmental degradations. Some gains in poverty reduction have stalled and worldwide inequality has grown. Periodic cycles of good and bad collective decisions are exemplified by the United States' withdrawl in 2017 from the Paris Agreement on climate change and then its re-entry 4 years later.
- The *orange sky scenario* is a precautionary tale of time running out with the sun setting on available options to maintain humanity's well-being and nature's benefits. The COVID-19 pandemic of 2020 has reduced worldwide economic activity and eased consumption of hydrocarbons, but in February 2022, the Russian-Ukranian war once again highlighted the world's strong and continued dependence on fossil fuels. Complexification of decision-making is exacerbated mainly due to the tension between energy security and the need for a climate conducive energy mix at all governance scales.

- The *white sky scenario* paints an image of the blue sky actually turning white due to misguided solar geoengineering attempts designed to offset atmospheric carbon dioxide rise. Biodiversity loss is irreversible in several biomes and its dampening effects on climate change are lost. Collapse of the Earth system becomes more likely with mass human migrations increasing due to food and water insecurity. Deep uncertaintity and unresolved risks dominate societies due to unpredictable and random decision-making.
- The *blue sky scenario* evokes a recovering planet and a more equatible global society that capitalises on the positive aspects of technology, including its sequencing and rate of change. Renewable energy has increased in both absolute and relative terms with respect to total energy consumption. Worldwide poverty reduction has rebounded from setbacks and continues to decrease synergistically with inequality. Wise collective decision-making prevails due to dynamic epistemic equilibria and the effective unity of processes and outcomes. A Sustainable Development Goal has been added to the original 17 to explicitly and comprehensively guide collective decision-making and taking.

All four scenarios leading to 2030 are discussed and compared in more detail below. Pair-wise contrasts can be used as qualitative measures of the starkest options among an even number of scenarios. Although there are no true likelihoods associated with scenarios in the sense that they are not forecasts or predictions, pair-wise comparison of relative scenario likelihoods can be used for qualitative risk assessments given that risks generally increase with negative likelihoods. As a first cut, scenarios can be categorised by whether they are possible, realisable, or merely desirable. Possible scenarios encompass all that are feasible; realisable scenarios are feasible scenarios given a set of specified constraints; and desirable scenarios are possible scenarios that may not necessarily be realisable or feasible [258]. Notably, exploring catastrophic climate change scenarios from bad-to-worst cases is said to be vital. The precautionary principle and prudent risk management demand that we thoroughly consider worse-case scenarios [253].

Setting the Scene: Approach and Selected Scenario Drivers

In our obscurity, in all this vastness, there is no hint that help will come from elsewhere to save us from ourselves—Carl Seagan

The essence of making the future visible is to understand the pattern of forces propelling the present into the future and to see the different places where those forces might lead. It is about thinking and asking the right questions to help detect the changes in the game in which you are a participant, whether they are related to where you live and work, or the interests that you have. It can extend to global issues like climate change, but it can also be about the societal trends in your own country or community that are bringing people together or driving them apart. It can be about technology trends in the time period that you have selected and what industries are becoming relevant or becoming obsolete [260, pp. 4–5]. And it is about your relationship with Nature. A suitable overall approach for democratic scenario development can be depicted by way of Fig. 9.2. The rationale for the four selected drivers of interest, namely, climate change, inequality, technology, and decision-making are discussed in more detail below.

In the years ahead under all scenarios, CO_2 and GHG policies will bring ongoing challenges as to how energy is produced, transported, and consumed given that energy systems contribute about three-quarters of the world's greenhouse gases. These challenges will extend to investments and strategies; to technology and infrastructure; and in relations among countries. The adaptability of established companies will be tested. New entrants will have to justify their business models. Dynamic partnerships and competition will characterise relationships among different types of companies. Energy security concerns will extend to the policy chains supporting low-carbon industries and to the minerals on which renewable energy technologies depend. Climate change will remain global, but action will be local—nations will respond in different ways based on the own particular circumstances. Developed countries will have more options and flexibility. Developing countries will struggle to balance their economic growth with decarbonising their economics. China's "9-dash map" or any of its successors concerning the South China Sea region will complicate international relations and global energy security [261, pp. 411–412, 144–149]. Finance flows will improve but still fall short of the levels needed to meet climate goals across all regions and sectors. Gross domestic product (GDP) will be increasingly challenged yet remain popular as the most suitable metric of societal progress in the absence of a widely accepted alternative measure of societal progress. Both the risks and benefits of artificial intelligence will remain a wild card but

Expert Driven	**Step 1: Select long-term drivers (rules of the game)**
	Example: climate change, procedural justice, world economy, inequality, technology, innovation
Citizen Driven	**Step 2: Identify 4 key opposing uncertainties**
	Example: sustainability or collapse / environment or economy (see Figure 9.1)
Citizen Driven	**Step 3: Creatively name the 4 scenarios from the opposing uncertaintity combinations and plot them on a quadrant**
	Example: "Orange Sky", "Grey Sky", "Blue Sky", "White Sky" (see Figure 9.1)
Expert Driven	**Step 4: Identify and monitor key indicators (flags) for each scenario or combination of scenarios**
	Example: SDG indicators, Planetary Boundaries, Human Development Index, Gini Coefficient, fiscal and energy budgets
Expert Driven	**Step 5: Analyse scenarios**
	Example: formally model time-bounded scenarios by the interplay of drivers and indicators and compare to relevant others by means of energy balances
Jointly Driven	**Step 6: Improve decision-making and decision-taking by diverse frames and methods**
	Example: assess future scenario and policy risks by subjective probabilites for final objective multi-criteria decision-taking

Fig. 9.2 Democratic scenario planning roles, steps, and examples. (Source: Author based on [260])

will continue to grow with international and national rules and regulations lagging behind.

Climate Change: Human-induced climate change beyond natural climate variability, characterised by more frequent and intense extreme events, has caused widespread adverse impacts and related losses and damages to both nature and people. Ecosystems and their biodiversity provision human livelihoods and provide ecosystem services. Although some development and adaptation efforts have reduced vulnerability across regions and sectors, the people and systems most at risk are disproportionately affected. The rise in climate and weather extremes has led to some irreversible consequences as human and natural systems are pushed beyond their ability to adapt. The level of acceptable risk depends on concurrent near-term trends in exposure, vulnerability, level of socio-economic development, and adaptation. Human society can adapt to or maladapt and mitigate climate change, while ecosystems can mitigate and adapt within limits. Climate-resilient development requires society and ecosystems to transition to a more resilient state. Global warming, reaching 1.5 °C in the near-term (2021–2040), would cause unavoidable increases in multiple climate induced hazards and present multiple risks to humans and ecosystems. Near-term actions that limit global warming to 1.5 °C, or near 1.5 °C, would substantially reduce projected losses and damages in human systems and ecosystems compared to higher warming levels but cannot eliminate them all**.** Taking action is enabled by good governance, adequate financing, capacity building, growing knowledge, and catalysing technology [262].

Inequality: Much of the economic inequality currently observed around the world is associated with rent seeking, the broad term that has come to mean capitalising on ownership without contributing to useful production. Yet no single class of ownership—whether private, government, or community—succeeds or fails on its own to impede or halt major resource deterioration, as shown for forests in multiple countries [56]. This suggests that inequality extends to the natural world by what has been called *impact inequality*. Estimates of humanity's total impact on nature show that we would require 1.6 Earths to maintain the world's present living standards [263, p. 1]. These living standards will become mostly aspirational if the worst dimension of economic inequality—inequality of opportunity which is both a cause and the consequence of inequality of outcomes—continues to suppress development as large numbers of individuals do not live up to their potential. Countries with high inequality tend to invest less in public goods, such as education, infrastructure, and technology, all of which could otherwise contribute to long-term economic prosperity. Reducing inequality has clear economic as well as social benefits. It strengthens people's sense that societies are fair thus enhancing social cohesion, making it more likely that more citizens live up to their potential and concurrently broadens support for economic initiatives. Government policies that aim for growth but ignore inequality are ultimately self-defeating, whereas policies that decrease inequality and boost employment and education improve the human capital that modern economies increasingly depend on [264, p. 288]. In the era of climate change, global inequality is a central concern for the fair allocation of burdens, costs, and desirable limits to growth across regions and time.

Technology: Changes in technology availability, accessibility, affordability, acceptability, and adaptability have transversal socio-economic and socio-environmental impacts. The term "4th Industrial Revolution" has been coined to describe the latest technological developments in cyber-physical systems such as: artificial intelligence, high-capacity connectivity, cloud computing, "big data", robotics, 3D printing, the "Internet of things", renewable energy systems, energy storage [265], smart electricity grids, and electric cars, among others. In addition, some previous technological trends continue their upward trajectories as is the case with information technology evidenced in part by the rise of social media and artificial intelligence. Quantum computing remains a wild card. And the biotechnology and materials revolutions continue to have strong momentum. Intense and sustained competition in the global marketplace creates incentives for rapid introduction and subsequent worldwide diffusion of new technologies. The technological ground rules are likely to change dramatically over the next thirty years with the potential for disruptive changes and significant impacts through employment displacement, lifestyle changes, and the globalisation of culture among others [47, pp. 368–369]. Governments have a crucial role to play in promoting appropriate new technologies beyond the provision of funds for national research and development. They set national policy objectives and budget priorities and play a vital role in determining market expectations. They also have key responsibilities for ensuring the flow of knowledge, investing in enabling infrastructure, and facilitating major demonstration projects [266].

Decision-Making: A trend towards the decentralisation of authority and greater individual autonomy is underway. Entities that have no formal authority, such as the Internet, persist and can give rise to other synergetic configurations as evidenced by blockchain technologies. In the public sector, individual autonomy can accelerate more open democracies and the devolution of authority to smaller and local units. Compounding these trends is the proliferation of new actors from citizen groups to global corporations that make governance an increasingly complex process [47, pp. 368–369], including the need for greater participatory decision-making. Decision-making and decision-taking are the main mechanisms of human agency by which humanity can respond to existing and emerging challenges, and by which it can advance a variety of possible solutions. Yet as the primary mechanisms of human agency, they are inadequately addressed as topics in the realm of sustainability. The 2022 article titled *Decision-Making for Managing Climate-Related Risks: Unpacking the Decision Process to Avoid "Trial-and-Error" Responses* provides a useful synopsis of climate-related decision-making tools, references, and case studies [267], but the democratic means, epistemic prerequisites, and decision-taking issues that the present study covers on and beyond climate change point to important gaps in the literature. In our evolutionary inspired schema, policy responsive scenarios are *jointly driven* by citizens and experts—citizens are like foxes and experts are like hedgehogs—both are needed in different measures. In bureaucracies and elsewhere, the task of information integration is insufficiently appreciated. Yet anyone at the top of an organisation, be it a president, prime minister, legislator, regulator, or a CEO, has to make decisions *as if* all aspects of a situation along with their interactions were being taken into account. Is it reasonable for the chief executive or a leader reaching down into the organisation for help to encounter only specialists and for integrated thinking to take place only when he or she makes the final intuitive judgements? [17, p. 346].

Grey Sky or Orange Sky?

> *Today we use 100 million barrels of oil every day. There are no politics to change that. There are no rules to keep that oil in the ground. So we can't save the world by playing by the rules*—Greta Thunberg

Grey Sky Scenario: In 2030, the sustainability clock stands at two and a half minutes to midnight with the strong prospect of accelerating decline. Six of the nine planetary boundaries have been exceeded. A tipping point of reversals can propel us into a chaos of catastrophes, or a critical mass of successes can propel us into a dynamic state of equilibrium. *Climate Action* (SDG 13; 5 targets, 8 indicators) has not benefited from precautionary conservation strategies for non-renewable resources. Temperatures of more than 2 °C above pre-industrial values have not been sustained on Earth's surface since before the Pleistocene Epoch [253]. Despite the growth in policies and laws addressing mitigation, interventions have been downscaled with focus mostly on adaptation, yet feasible and low-cost options for mitigation and adaptation are available with more synergies than negative trade-offs with the SDGs. System transitions call for a significant upscaling of a wide portfolio of mitigation and adaptation options. Hard and soft limits to adaptation have been reached in some ecosystems and regions at the same time than maladaptation is happening in other sectors and regions. If warming exceeds a specified level, it could gradually be reduced again by achieving and sustaining net negative global CO_2 emissions. Near-term actions require high up-front investments and involve potentially disruptive changes which can be lessened by a range of enabling policies. Some future changes become unavoidable but can be limited by deep, rapid, and sustained global greenhouse gas emissions reduction. Rapid and far-reaching transitions across all systems and sectors are necessary to achieve deep and sustained emissions reductions to secure a liveable future [268].

Under a grey sky, the *Reduced Inequalities* goal (SDG 10; 10 targets, 14 indicators) shows mixed results. Nearly half of all carbon emissions since the dawn of industrialisation have benefited the development of just 10% of the current global population [151, p. 141]. There is widespread recognition that poverty is one of the most pernicious effects of inequality. Over 800 million people are still suffering yet the solutions are known—peace, schooling, universal basic healthcare, electricity, clean water, toilets, contraceptives, and microcredit to start or strengthen market forces. There is no innovation needed to end poverty—all that is needed is walking the last mile with what has worked everywhere else. People who have remained in extreme poverty are likely to continue the trend of having large families thereby offsetting the benefits of available goods and services necessary for a decent life. Quickly providing these necessities to the final billion of the world population is a widely accepted fact-based priority [269, pp. 240–241] that is not being implemented sustainably. Training provision versus technological innovation continues to be hampered by skewed labour supply and demand confounding wage inequalities. Coarse-grained measurements like country capital/income ratios prevail [270, pp. 63–65, 382–386].

In the grey sky scenario, *Industry, Innovation and Infrastructure* (SDG 9; 8 targets, 12 indicators) is primarily driven by national interests and unhealthy competition. Renewed worldwide energy efficiency policies become increasingly necessary to avoid lock-in of outdated technologies. Lock-in of fossil fuels prevails and continues to be subsidised. This is exemplified by the capture of the international COP process by the fossil fuel industry. Overall, sustainable development technologies have not been prioritised, nor coordinated, but some progress is being made. In the electricity sector, demand-side interventions like installation of heat pumps in public buildings are a codified requirement in some countries, but incentives for households and the broader private sector are lagging. Data centres consume an estimated 1% of the electricity produced globally, which is somewhat ameliorated by cloud and edge computing systems, but within the context of a growing demand for data services. Artificial intelligence and "big data" have yet to produce game changing results. A low-tech strategy of keeping hydrocarbons in place has immediate benefits whether it is coal in the ground or forests in the Amazon, but this low-tech solution is dependent on effective high-tech monitoring. Managing the decarbonisation of economies by carbon capture and storage has proved to be impractical and uneconomic. The circular economy both as economic policy and practice is nascent. Recycling of all of the world's money supply is misaligned with the life cycles of all of its goods and services.

Under a grey sky, periodic cycles of good and bad collective decisions proliferate exemplified by the United States' withdrawl in 2017 from the Paris Agreement on climate change and then its re-entry 4 years later. Large temporal gaps between decision-making and decision-taking regarding many sustainability issues hamper smooth transitions. Geopolitical issues punctuate and draw attention away from urgent sustainability concerns. A lack of visionary leadership exacerbates the prevailing lack of urgency. Some countries add or pass new laws and regulations to protect the environment, but most do not. A beneficial "edge of chaos" equilibrium between economic and enivironmental cosiderations has not been achieved. Sustainability best practices are not widely shared or effectively communicated among stakeholders. The state space of sustainability knowledge and decision-making has more "knowns" than "unknowns", and relevant ergodic-breaking approaches to improve decision-taking are under investigation. Global decision-making is consensus dependent yet remains parochially driven. Generative decision-making by AI systems continues to be poorly understood. Al Gore's seven-point yes/no decision checklist is not used widely by civil society and activists to effectively target fossil fuel companies and their funders to mitigate climate change

[271][1]. Some sustainability indicators are moving in the right direction; others are moving in the wrong direction.

Orange Sky Scenario: In 2030, the sustainability clock continues its downward trend and stands at 2 min to midnight. Six of the nine planetary boundaries have been exceeded and downward trends are accelerating. *Climate Action* has taken a turn for the worse. Under voluntary national commitments for climate action, greenhouse gas emissions are projected to rise globally by 14% by 2030 [70]. This trajectory puts the world on track for a temperature rise between 2.1 °C and 3.9 °C by 2100. If all nationally determined contributions (NDCs) for climate change reductions from the 2015 Paris agreement on climate change are fully implemented by 2030, average warming of 2.4 °C is expected by 2100. Meeting all long-term pledges and targets could reduce this to 2.1 °C. These estimates, however, lead to dangerous pathways for the Earth system. Global greenhouse gas emissions as estimated by NDCs make it likely that warming will exceed 1.5 °C during the twenty-first century making it harder to limit warming below 2 °C. Although climate adaptation planning and implementation have progressed across all regions and sectors, adaptation gaps exist and continue to grow. With increasing global warming, losses and damages increase and additional societal and natural systems reach their adaptation limits [268]. Despite three decades of efforts and some progress under the United Nations Framework Convention on Climate Change, greenhouse gas emissions continue to increase. There is a rapidly closing window of opportunity to secure a liveable and sustainable future for humanity.

Under an orange sky, global North versus South inequalities grow jeopardising SDG 10. A "pay" component by governments and companies is seriously opposed and sabotaged by powerful special interests. The International Energy Agency (IEA) estimates that another $20 trillion of oil, gas, and coal development is literally in the pipeline through 2040 [151, p. 136]. Globally, a third of oil reserves, half of gas reserves, and over 80% of coal reserves should remain unused through 2050 to meet the less ambitious target of a 2^0 C temperature rise [272]. There is sufficient global capital to close investment gaps but barriers to redirect capital to climate action are not being overcome. Both adaptation and mitigation financing fall short by several orders of magnitude. Global financial flows for adaptation are insufficient and constrain implementation options, especially in developing countries [268]. The Genuine Progress Indicator (GNI) is consistently overlooked by multilateral organisations as an alternative or complementary measurement to GDP. The circular economy is losing proponents and influence as an alternative economic model. Recycling of all of the world's money supply is increasingly misaligned with the life cycles of all of its goods and services.

---

[1] (i) Does the company have real net-zero commitments, a genuine phase-down plan; (ii) Are they committed to full disclosure? (iii) Are they going to spend windfall profits on the transition? (iv) Are they committed to transparency? (v) Will they end their anti-climate lobbying? (vi) Will they end their greenwashing? and (vii) Will they be in favour of reforming the COP process so that the petro states don't have an absolute veto on anything the world wants to discuss or act on?

Under an orange sky, cross-linkages of SDG 9 to the other SDGs become weaker. Although economically competitive, renewable energy has not substantially displaced fossil fuels as the primary source of energy in most economies. The fossil fuel industry continues to be heavily subsidised. Governments around the world continue to subsidise with taxpayer money the fossil fuel industry to the tune of more than a trillion dollars, which is five times larger than the amount in 2020 [271]. Energy efficiency gains have stalled. Millions of people continue to lack electricity, and the deployment of clean cooking technologies is slow in the developing world jeopardising their populations' health [119]. Proponents of AI regulation have an ulterior motive for requesting government regulation, namely, to create market barriers to entry for less developed competitors. NASA's Technology Readiness Level scale is not being appropriately adapted by technology developers to set research priorities and to support innovation programmes [273].

In the Orange Sky scenario, collective decision-making is in the realm of complexification with paradoxes, tensions, fragmentation, and uncertainties dominating possible sustainability solutions. A beneficial "edge of chaos" equilibrium between economic and enivironmental cosiderations has become even more difficult to attain. The state space of sustainability knowledge and decision-making has an inordinate number of "unknowns" compared to "knowns", and relevant ergodic-breaking approaches to improve decision-taking are not recognised as such. Large temporal gaps between decision-making and decision-taking are frozen in place in many sustainability areas. Swift action before the sun sets on the available options can make the problem smaller. Evidence-driven decisions are disadvantaged by patchy sustainability data from participating countries. Global decision-making continues to be consensus dependent and parochially driven. The automated decision-making of AI is paradoxically shrouded in unknowns and on its way to becoming unknowable due to its rapid progress and lack of international regulations making regulatory capture more likely. Intergenerational driven sustainability decision criteria are discounted under rapidly closing windows of opportunity. Al Gore's seven-point yes/no checklist has fallen into disuse by civil society and activists reducing pressure on fossil fuel companies and their funders to mitigate climate change. Some indicators are moving in the right direction; most are moving in the wrong direction.

White Sky or Blue Sky?

Attempting to charge real costs is the principal element of the required economic transition from living in large part on nature's capital to living mainly on nature's income—Murray Gell-Mann

White Sky Scenario: In 2030, the sustainability clock stands at half a minute to midnight—its lowest reading. Nine of the nine planetary boundaries have just been exceeded. The world is at the brink of nuclear war due to the resultant geopolitical tensions arising from Iran's acquisition of nuclear weapons, continued North versus South Korea aggression, repercussions in Europe from the Russian-Ukrainian war, the China-Taiwan stand-off, as well as the unresolved Palestinian question. Pandemics surge due to multiple systemic crises. Viruses have come back to life from melting permafrost and constitute an unknown global health hazard. Climatic and non-climatic risks increasingly interact, creating compound and cascading risks that are more complex and difficult to manage. Projected CO_2 emissions from past fossil fuel infrastructure without additional abatement exceed the remaining carbon budget for the 1.5 °C target. Every increment of global warming intensifies multiple and concurrent hazards. Delayed mitigation and adaptation action locks-in high-emissions infrastructure, raises risks of stranded assets and cost-escalation, reduces feasibility, and increases losses and damages [268]. The lack of an international agreement banning atmospheric aerosol injection raises the potential of catastrophic risk from the sudden failure of the injection system. This shifts the risk distribution such that the median outcome is better than the climate change it is offsetting, that is, the high-impact, low-probability risk that it has created is worse than the warming it was trying to solve [253, 274]. The Sixth Mass Extinction is now irreversible.

Under a White Sky, higher population growth rates become an even more potent threat multiplier. The cost-of-living crisis has exacerbated inequality across the world increasing forced migration and related political tensions. Rising energy, commodity, and shipping prices have increased the cost of producing renewables adding uncertainty to energy transitions and sustainable development. High inflation and surging public debt levels hamper countries' ability to support vulnerable groups. A global recession is raging due to a runaway mortgage financial crisis. Climate change risks have drastically reduced the value of homes due to the unavailability of home insurance as required by banks to extend mortgages [275]. Home ownership is becoming an ever more distant dream for those less well off. An increasing number of leading global banks exit the Net-Zero Banking Alliance [276] undermining climate change goals and the underlying United Nations Environment Programme (UNEP) finance initiative. 3BL reporting by large companies has stagnated and rating agencies do not consider any of the available "green" factors in their credit ratings of countries. GDP is unchallenged as the most suitable metric of societal progress, while GNI falls into disuse from a previously growing base. The circular economy is no longer an aspirational goal. Recycling of all of the world's money supply is almost completely misaligned with the life cycles of all of its goods and services. Disruptions to education from the COVID-19 pandemic and the disproportionate adverse impacts on low-income households continue to worsen intergenerational mobility [119].

In the White Sky scenario, the unintended consequences of technology proliferate due to the lack of appropriate initial sequencing and subsequent responses to negative outcomes. Technological solutions to environmental problems are more expensive than the preventative measures to avoid creating the problem in the first place [141, pp. 504–505]. Anthropogenic mass on Earth in the form of concrete, aggregates, bricks, and asphalt continues on an upward trajectory in relation to natural biomass in the form of trees and shrubs, while the global mass of plastic is increasing relative to the overall mass of all terrestrial and marine animals combined [277]. Most new technologies create cascading unintended consequences. Earth system dynamics remain inadequately modelled with persistent complexity and high degrees of uncertainty undermining effective policy directions. The greater and faster penetration of the Internet has increased cyber insecurity. AI is permanently disruptive after reaching a cost-benefit tipping point.

Under a White Sky, uncertainty and risk dominate societies exacerbated by poor leadership confounding decision-making and decision-taking. The state space of sustainability knowledge and decision-making has more "unknownables" than "knownables", including relevant ergodic-breaking approaches to improve decision-taking. Larger temporal gaps between decision-making and decision-taking regarding many sustainability issues overwhelm smooth transitions. A beneficial "edge of chaos" equilibrium between economic and enivironmental considerations becomes impossible to attain. The UN's COP process has lost credibility and is indefinitely suspended together with the 2015–2030 Sendai Framework for Disaster Risk Reduction. The rise of the BRICS formation (Brazil, Russia, India, China, South Africa, plus new entrants) is disruptive in geopoltics with unforseeable consequences for the Earth System. Existing coperation among Russia, China, and Brazil is in tension with the needed cooperation with the United States, Canada, and Australia—the six countries with most of the potential for tree restoration to combat climate change [278]. Climate targets are being re-negotiated, weakening them rather than strengthening them due to unresolved global North-South differences, untested international coalitions, and a lack of attention on the collective decision-making process itself. All indicators are moving in the wrong direction.

Blue Sky Scenario: In 2030, the sustainability clock has been wound back to three and a half minutes to midnight. Five of the nine planetary boundaries have been exceeded. Greenhouse gas emissions are no longer increasing putting the world on track to reach a temperature rise of 1.5–2^0 C relative to pre-industrial levels. Many adaptation options persist and are being increasingly used to manage projected climate change impacts, but their continued implementation depends on the effectiveness and adequate capacities of decision-making and governance systems. These and other enabling conditions similarly provide impetus for climate-resilient development. Resilient development is enabled by governments, multilateral organisations, civil society, and the private sector who make inclusive development choices that are integrated across governance levels, sectors, and timeframes. Risk reduction, decision-making processes, as well as equity and justice are prioritised. Climate-resilient development is facilitated by international cooperation and by developing partnerships with traditionally marginalised groups, including women, youth, ethnic minorities, local communities, and indigenous peoples. These partnerships are effective because they are supported by constructive political leadership, institutions, resources, including finance, as well as climate services, information, and decision support tools [262, pp. 22, 31]. Energy intensity as a measure of efficiency at the country level is replaced by measures closer to consumers, in particular, useful exergy over final energy ratios. The *Cape Town Global Action Plan for Sustainable Development Data* is being rigorously implemented. Climate action is better enabled by political commitment, well-aligned multilevel governance, effective institutional frameworks, laws, policies, and strategies, as well as enhanced access to finance and technology. Clear goals and targets, integration across multiple policy domains, and inclusive governance processes improve effective climate action. Reductions of greenhouse emissions work to support climate resilience and are applied widely through regulatory and economic instruments. Climate-resilient development is enhanced by drawing on diverse knowledge and benefitted from prioritising equity, climate justice, social justice, inclusion, and just transition processes [268].

Growing green financing has progressively helped to reduce Global North versus Global South inequalities putting SDG 10 within reach. Sustainable development has been fully embraced by the world business community and the major fossil fuel companies have stopped denying that they have been major contributors to the Earth's climate crisis. Subsidies to this industry have been eliminated by most countries. Extreme income inequalities have been reduced within and across all countries as measured by Palma ratios[2] [264, pp. 288–291]. The scaling down of USAID has in part been counterbalanced by growing non-profit organisations such as *GiveDirectly*,[3] which has maintained an online platform to combat poverty by enabling individuals to donate cash directly to recipients. Developed countries have implemented a universal basic income for their citizens based on the Spanish model and wealth autocrats are paying their fair share of taxes. National government budgets earmarked for clean energy transitions or monies encumbered for sustainability initiatives are systematically monitored and aggregated at multiple intermediate levels and subsequently consolidated into a global financial sustainability ledger. 3BL reporting has become an international accounting standard and is used to assess the credit scores of multinational and large companies and transparently reported on in company financial statements. Rating agencies have incorporated green financing pledges as line items for determining the credit worthiness of countries. GNI has become the dominant measure of societal progress and comparison.

Under a Blue Sky, *Industry, Innovation, and Infrastructure* progress is led by renewable energy which surpasses 50% of installed capacity with floating solar panels generating electricity while reducing evaporation in water scare areas. Graphene-based batteries increasingly store electricity at scale, and graphene composite transmission lines distribute electricity efficiently across large geographical regions. Thorium-based modular nuclear reactors have reached technological maturity and are used widely to power electric vehicle charging stations with excess electricity

[2] The top 10% of a nation's foreign and domestic post-tax income divided by its population's poorest 40% post-transfer income.

[3] https://www.givedirectly.org

being fed back into smart electricity grids. Strategically placed modular nuclear reactors power desalination plants. The Advanced Research Projects Agency-Energy of the United States has successfully promoted nuclear fusion projects and both the private and public sectors are moving potentially closer to inexhaustible clean energy. Carbon capture, storage, and use is increasingly carried out at scale using basalt dust from quarrying waste to remove atmospheric carbon with the added benefit of its use to fertilise cropland. Open-source energy modelling software has become progressively accurate and serves a dual purpose: for national energy security planning, as well as a tool to support ongoing global climate mitigation and adaptation. In transportation, sustainable aviation fuel production exceeds 300 million litres coupled with a 20% increase in demand, which requires the creation of about 300 new biomass and CO_2 fuel plants. Syngas, a mixture of carbon monoxide and hydrogen, can be prepared from biomass or wastes, and increasingly, from captured CO_2 and green hydrogen using renewable energy.[4] An increasing number of airlines have committed to using sustainable aviation fuels [279]. Hydrogen-based fuels from CO_2 have achieved 80% market penetration aided by onsite production methods. Electric Vertical Take-Off and Landing (eVTOL) vehicles are being increasingly used as both public and private transportation. The circular economy both as economic policy and practice is embraced by the majority of stakeholders. Recycling of all of the world's money supply is closely aligned with the life cycles of all of its goods and services.

Decision-making under a *Blue Sky* is marked by wise collective decision-making setting humanity on a sustainable path with the long-term horizon and future generations in mind. The state space of sustainability knowledge and decision-making has more "knownables" than "unknownables", and relevant ergodic-breaking approaches to improve decision-taking are being implemented. Temporal gaps between decision-making and decision-taking regarding many sustainability issues have narrowed making transitions smoother. Climate-resilient development advances sustainable development for all. Increased international cooperation has gained strength and is characterised by inclusive governance and coordinated policies, including improved access to adequate financial resources for vulnerable regions, sectors, and groups [268]. A beneficial "edge of chaos" equilibrium between economic and enivironmental cosiderations has been gained and maintained. The United Nations has coordinated the adoption of a new treaty setting out the criteria under which SDGs and successor targets will be developed. This facilitates North-South country negotiations and minimises future disputes by parties being able to invoke this standing framework. A new agency—*UN Energy*—has been established and funded for the coordination, integration, and adjudication of energy issues across regions and nation-states. UN Energy is tasked with drafting model sustainability related laws and regulations for possible modification and adoption by nation-states. The foundational importance of collective decision-making processes

[4] The American Society of Testing and Materials has approved nine fuels for blending at a ratio of up to 50% with conventional petroleum-based jet fuel.

has been acknowledged and elevated to be an SDG goal with appropriate targets and indicators. Improved collective decision-making facilitates the domestication of sustainability-related treaties, compacts, and agreements into national constitutions, which are then mirrored in subordinate jurisdictions. The massive societal reuse model of cognition has replaced the old System 1 and System 2 models, and the knowledge and decision-making state space of sustainability has been largely mapped providing better conceptual guidance. The knowledge base on the size and composition of decision-making and decision-taking groups is being continuously curated and updated. Operational paradigms have been re-imagined by organisational working groups and guide collective decision-making structures, processes, and tools. New decision modelling software has been created or successfully redesigned, and the automated decision-making of AI is made transparent creating appropriate entry points to correct any emerging or cascading biases and incorrect information. Participatory decision-making is democratically driven by directed epistemic attention and is widespread with multiple improving methods and results. All indicators are moving in the right direction.

Energy Balances: A Proposed Scenario Analysis Method

Now is the time to understand more, so that we may fear less—Marie Curie

The importance of what to measure and how to measure it can be determinative in guiding thought and action, including in planning scenario analysis. Information about the availability of energy and money and their distribution are two simple yet powerful coarse-grained means of measuring sustainability. Given that energy cannot be created or destroyed according to the first law of thermodynamics, each planning scenario under analysis can be quantified in part by estimating its particular *energy balance* (budget). For example, the International Energy Agency (IEA) routinely produces energy balances for its member country's energy systems which can be depicted as Sankey Diagrams. Energy flows can serve as a transversal metric for both natural and man-made systems. Demand for energy is a major driver of climate change; therefore, decarbonising national energy systems is a necessary condition for rebalancing energy demand and supply, as well as in indirectly mitigating biodiversity loss [263, p. 3]. Consider a global *sustainability energy flow ledger* at different levels of aggregation that can be monitored as close to real time as possible and as readily understood by the general public as the term "GDP".[5] Global energy use over the past century has typically exhibited 2–3% growth per year, but physics puts a hard limit on how long our energy growth could possible continue [280]. The IEA's snapshot of the world's energy flow for a given year, unlike a detailed general ledger, provides analysis of energy production compared to energy consumption,

[5] Gross domestic product (GDP) is not a sustainability friendly metric, see, for example, the monographs *Measuring What Counts* and *the Progress Illusion.*

but minimally against energy losses.[6] Yet according to the Lawrence Livermore National Laboratory of the U.S. Department of Energy, the world's energy flow exhibited more energy losses (rejected energy/waste heat) than actual energy services in 2011. In the United States, "rejected energy" continued to outstrip energy services in 2021 perpetuating the so-called *primary energy fallacy* [281], that is, completely discounting the losses in the energy consumption ledger.[7] This is a prime example of the physics that speaks to the issues of efficiency and conservation that can inform both public policy and personal decision-making. In short, energy efficiency as currently measured by energy intensity (units of energy per units of national GDP) will prolong the rebound effect known as Jevon's Paradox if targeted mitigation measures are not planned for.

Policymakers more often than not regard economic growth as a proxy for human development and social progress. Past a certain point, however, the correlation between GDP and social indicators breaks down. Strong evidence reveals a robust relationship between GDP and energy use; however, modelling suggests that with existing growth trajectories, absolute reductions in energy use are unlikely to be achieved. Existing climate scenarios maintain a significant disparity in per capita energy use between the Global North and Global South but without adequate consideration of a counterfactual, that is, that high-income nations can maintain economic stability, invest in innovation, and achieve strong social outcomes without the need for additional growth. The main way to achieve the Paris climate goals is to significantly reduce energy demand, making it easier to accomplish rapid decarbonisation [282]. To break the cycle at the Earth system level, a *reduction* in global human energy consumption as measured by *useful exergy* (energy that can produce useful work and services) relative to *increases* in energy consumption by Earth's ecosystems, measured by *carbon sinks*, whether biological or physical, is needed. This is the core socio-biophysical balance that has been destabilised by unchecked economic growth. In the field of ecological economics, global level flows of ecological services are considered as economic valuations [283] calculated in dollars rather than terajoules, but separating the economy from energy and material flows fails any scientific scrutiny yet dominates the practice and teaching of economics [151, p. 191]. Our academic institutions could be put to more impactful use by contributing to the definition of a future world that openly acknowledges thermodynamic limits and respects planetary limits [280]. For example, measurement of socio-economic metabolism by *ecological footprint analysis* can increase the visibility of humanity's impact on the environment. And despite money being ubiquitous in modern societies, it is not utilised constructively as a broad sustainability

---

[6] "Useful exergy" is the more technical and accurate term for energy services consumption, exergy being the actual capability (versus energy's potential capacity) to deliver energy services, such as lighting, cooking, and heating. It is also more accurate to state 180 countries and regions instead of "world".

[7] 290,000 petajoules of rejected energy, mostly due to unproductive heat lost to the environment, versus 210,000 petajoules of delivered energy services. More Sankey diagrams are available at https://flowcharts.llnl.gov/commodities/energy

metric. Interestingly, in areas where electricity is available and used, it can serve as a composite indicator that brings energy and money together (given established systems of billing and other instruments of financial accounting) with the added benefit that standard incentives/disincentives can be brought to bear to meet policy objectives. It is also critical to embed risk measurement as a constituent part of other composite metrics, for example, by the use of "jumping indicators". Joseph Stiglitz recommends working towards the development of a statistical system that complements measures of market activity by ones centred on people's well-being, as well as measures of sustainability [284].

Ranking and choosing planning scenarios to inform policy in terms of *comparative energy savings* can serve as a complementary measure to market activity. Most notably, GDP does not account for the depreciation of assets, including those of the natural environment. By measuring *inclusive wealth*, however, a concept that encompasses natural assets, more sustainable growth can be encouraged while taking into account future generations. Introducing natural capital into national accounting systems would be essential for making inclusive growth our measure of progress. Investment in physical accounts and valuation would enhance data and modelling approaches, as well as the quality of capital accounts for purposes of improving decision-making at scale around the world [263, p. 4]. A 2018 International Monetary Fund (IMF) book recommends a *System of National Accounts* (SNA) for the measurement of SDGs which incorporates but goes well beyond GDP. Its central feature is a conceptual "Super Balance Sheet" with broad asset classifications that include durable consumer goods used to produce services, human capital, and access to resources, such as clean water, clean air, education, health, and infrastructure so as to produce an expanded household net worth. The growth of data availability with digital and geospatial information makes it possible to construct such Super Balance Sheets. However, the unit of measurement is limited to the household since it is not feasible to allocate the ownership of environmental assets across institutional units. Therefore, the development of Super Balance Sheets for other sectors would be needed and would require an allocation of environmental assets to the units and sectors with a direct benefit. By the IMF authors' own observation, the SNA framework is less well known than GDP and more difficult to interpret. But the most welcome aspect of the SNA initiative may be its provision for a satellite accounting framework which allows for conceptual variations from its central framework [285]. Such (and other) conceptual variations may be seeded by thinking of money as coarse-grained human inputs and energy saving changes as coarse-grained Earth outputs.

Chapter 10
Who Decides for Whom and Why?

The penalty that good men pay for not being interested in politics is to be governed by men worse than themselves—Plato

Earth system governance as represented by the 17 SDGs has required integrated, holistic, and coherent policymaking in which decision-making and decision-taking involve actors from the public and private sectors, as well as civil society. Such governance approaches can be decisive when dealing with highly dynamic and long-term sustainability problems. This in turn requires the synergies of institutions and procedures, a governance characteristic that often seems to contradict traditional rationalist problem-solving approaches. Scholars maintain that both representative and direct democratic governance structures may facilitate the decision-making process, implementation, and acceptance of policies directed towards the achievement of sustainable development [190]; however, representative democracy can become unwieldy when multifaceted legislation and its administration are needed [286] making direct democracy more suitable for such matters as single issue referenda. Deliberative democracy and participatory democracy (yet other classes of polities) are in the mid-range between the former two, but all are encompassed in the new metaphor of an open epistemic democracy. An important difference between participatory democracy and open epistemic democracy is that the former is based on discrete and time-bounded participatory methods, whereas the latter relies on continuous learning and open-ended participation. Furthermore, by seeking to balance citizen versus expert inputs in the design and sequencing of open epistemic participation, an "epistocracy" can be forfended. Some of the enabling conditions to promote an epistemic democracy have been well paraphrased by Josiah Ober to include:

1. The specialised knowledge possessed by multiple sophisticated individuals is brought into the solution space; irrelevant information and expertise are excluded from that space; and each relevant knowledge input is given appropriate weight in the final policy;
2. No fixed and limited set of experts will possess the range of knowledge necessary for addressing the multiplicity of problems that will confront a state (or other organisations) over time; and.

E. Bellos, *Sustainability Thinking*, https://doi.org/10.1007/978-3-032-11244-6_10

3. Adapting the aggregated knowledge relevant in one solution space to different but relevantly similar problems in another space will speed the process of innovating new solutions [186, pp. 118–147].

Democracy is based on the assumption that the mêtis of its citizenry should, in mediated form, continually modify the laws and policies that apply to them. Mêtis is that vast space between the realm of genius, to which no formula can apply, and the realm of codified knowledge, which can be learned by rote. Innovations by mêtis typically involve the recombination of existing elements but without the restrictions of *techne* and *episteme* which cannot on their own fully explain scientific invention and insight. The rules of techne and episteme are the specification of how knowledge is to be verified, expressed, and codified, *once* it has been discovered by mêtis [8, pp. 316–324]. Thus, our stated objective of deriving the most benefit from citizen diversity while drawing on the cognitive resources of subject matter experts commends open epistemic democracy as the mechanism to achieve the requisite dynamic equilibria of *citizen experts* in relation to *expert citizens* (both groups are citizens after all). From the global existential risk and metasophia perspectives, the unity of participatory and representative democratic structures with those of commensurate epistemic processes that are variously nested in holon-like configurations increases the likelihood of attaining sustainability and other important societal goals. The decision-making authority in such an architecture is derived from the moral power bestowed by its people on their duly empowered representative bodies in their various transnational or devolved forms or by the legitimacy bestowed upon duly empowered representative decision-making groups to achieve a common goal by virtue of the shared intentionality of a community. And in a framework of known roles, decision-making is the final stage of collective deliberation followed by decision-taking as the final stage of its implementation. Collective decision-making would not be possible without the directed epistemic attention needed for clarity *of* purpose, and then recursively, as the needed clarity *for* purpose. Moreover, since decision-makers do not have the entirety of means to implement their own decisions, practical decision-taking would not be possible without bespoke programmes and projects guided by adaptive theories and methodologies[1] so as to be more context and risk responsive.

The Collective/Democratic Decision Model

Progress is impossible without change; and those who cannot change their minds cannot change anything—George Bernard Shaw

The proposed Collective/Democratic decision model presented below aims to improve the requisite epistemic balance between well-meaning citizens and cooperating subject matter experts to help reach goals at all system levels. It is rooted in

[1] Adaptive programmes = adaptive theories of change + adaptive theories of action + adaptive theories of measurement + adaptive risk management + adaptive financing + outlier monitoring + evaluative learning.

Robert Goodin's "all affected principle" which holds that all those possibly affected by the outcome of a decision ought to have a voice in it, and as such, this principle may be considered to be a moral *first* concerning the just way to extend rights and responsibilities in a democratic regime [287]. Both conceptual and computational models can serve as stand-alone schemas or as tools for the refinement of higher-order schemas and theories, as well as other analysis techniques. In his book, *The Model Thinker*, Scott Page details no less than twenty-six types of models ranging from modelling human actors to systems dynamics. Becoming a "many-model thinker", however, calls for creatively adjusting assumptions and constructing novel analogies to apply a model developed for one purpose to a new domain or context, which demands more than mathematical competence—it requires creativity [32, p. 40]. A quality check on the usefulness of models has to do primarily with their relevance. A model based on an irrelevant narrative is much worse than a model based on a wrong formalism. Such a model will keep providing irrelevant and misleading conclusions for ever [288, pp. 129–130]. The role that models play in organised knowledge production is illustrated in Fig. 10.1.

We have learned from social psychology that human action can be influenced by an idea known as the ideomotor effect. Although priming occurs at the subconscious level, it can be elevated by language as the idea jumps by repeated leaps along the speech hierarchy together with the intention behind the meaning and the train of thought [10, p. 210]. Consider that when primed with stereotypes of elderly people we become more conservative in our opinions, while those primed with politician stereotypes become more verbose [5, p. 167]. Primed ideas have the ability to influence other concepts and extend to political science in that they have been shown to influence voting behaviour. Effects may not seem highly consequential at the individual level, but even a few percentage points arising from changed minds can tip an election [85, pp. 52–58]. The lack of timely and trustworthy decision-taking may in practice undermine or even negate the original controlling decisions. Notice that the Sixth Assessment Report of the Intergovernmental Panel on Climate Change (IPCC, 2022) makes the point that many published studies on the usefulness of decision and analytic methods in managing climate risks are theoretical rather than practical, making it difficult to find assessment methods on which to base final decisions. Bayesian, Deep Uncertainty, and Elicitation methodologies have been the most readily available across decision-making contexts, while more specific method classes tend to be more oriented towards government processes. This finding highlights a key gap in the extant literature that calls for real-world experiences to be published and mapped according to their utility for different tasks, thereby creating

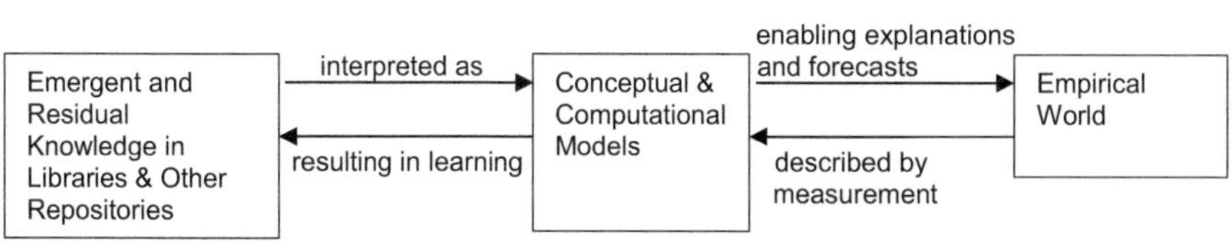

Fig. 10.1 The role of models, measurement, and learning for organised knowledge production. (Source: Author based on [289])

a library for policymakers to identify suitable resources, such as those in emerging communities of practice of decision practitioners.

One of the class of problems that the Collective/Democratic Decision Model seeks to address in a greater context is "group-think", a concept and term coined by Irving Janis to explain the John Kennedy's administration's 1961 failure in the Cuban Bay of Pigs invasion. More specifically, Janis identified eight characteristics that lead to group-think: self-assessments of invulnerability; the discounting of warnings and self-protection rationalisations; the perpetuation of stereotypes; self-censorship; cohesion of in-group morality; pressure on dissenting voices; the adoption of "mindguard" roles; and a delusion of unanimity [112, pp. 107–108]. Paradoxically, as powerful as these findings have been, one of the motivations for writing this book is that after more than sixty years the conception of group-think continues to be the reference case without any further generally accepted insights as to what can go wrong in group decision-making. Yet ways and means can be devised to improve group effectiveness. For example, people who need to collaborate with each other should learn, train, feel, and engage in rituals together—in person, at the same time [13, pp. 224–226]. From team assignments and activities to acknowledging and elaborating on member contributions, creativity would also be needed to design the most promising interventions and techniques for a particular context and time. Ultimately, when we gain a theory of mind, we also gain the ability to imagine what it must be like to think like another person, to see and experience things as they do, to have different opinions because of different exposures to different events. This is analogic perspective taking, a higher form of cognition that takes a lot of effort and is consequently rarely undertaken unless affirmatively called for [179, p. 50].

To get to doing from thinking, we can be guided by Hong and Page's Cognitive Model of collective wisdom with its emphasis on the "barbell" of diversity and sophistication. To do so, the context in which collective decisions are taken is foundational to better outcomes. It begins with considering the causes of diversity and sophistication which are many. Diversity may flow from identity, as well as different sources of experience and information. Sophistication, however, is dependent on attention, motivation, experience, and information. The common denominator of this "barbell" is information, in other words, epistemic considerations. It is important to also keep in mind that increasing individual-level sophistication can have costs: decreases in diversity if too great can offset increases in individual sophistication [152]. Most groups are better at reaching solutions through the clash of arguments than through joint deductive logic. Groups think best when they allow competing perspectives and contentions to take shape, be aired, and then be combined, recombined, and evaluated. This way of thinking is non-linear by first expanding the space of options and possibilities and then "closing them down". We may use wide networks to gather information about options and then use a small organisational group or individuals to make decisions under the uncertainty inherent in the future. Understanding better how focus is managed may turn out to be key to understanding collective intelligence and wisdom and how a dynamic balance should be achieved between the wide peripheral vision and directed attention needed to address a multitude of informational signals along with the motivation required for action. Comprehensive understanding takes much more time and energy than correct communication. In unstable environments where any item of knowledge is likely to be incomplete, groups with more capacity to reimagine their categories and thinking modes may adapt better. In principle, any group should optimise for stable environments with a well-thought-out division of labour until signs of change appear, at which point they should devote more to environmental scanning, re-reflecting on strategies and options, and mobilising resources that can be redirected to address opportunities or threats [24, pp. 75, 96]. This is the intuition behind the Collective/Democratic Decision Model which also capitalises on the benefits of diversity and expertise while pointing to some specific methods by which to achieve these objectives. But it is inherently impossible to know (except in retrospect) what the best dynamic balance should be for collective wisdom to materialise over time. In practice, a "ladder" from neuroscience and psychology to political science and social science would be needed to establish new directions of research to improve group decision-making and decision-taking.

The UK Design Council has developed a visual representation dubbed the "Double Diamond" fashioned to be applicable in any field to capture any design and innovation process regardless of specific methods and tools. The first phase has been described as *designing the right thing* and consists of two periods, one divergent named "discover" and then one convergent named "define". The second phase can be summarised as *designing the thing right* and is also made up of two periods, one divergent named "develop", and one convergent named "deliver" [290]. This flexible and useful tool can be adapted to both develop and depict the collective *decision-making processes* and *decision-taking outcomes* needed for sustainability.

The Collective/Democratic Decision Model depicted below is largely discrete in shorter time horizons but becomes continuous in longer time horizons. The left segment represents the decision-making inputs of engaged citizens or group participants, and the right segment the decision-taking selections of empowered experts or thought champions. The first segment and node in Fig. 10.2 marks initial decision-making based on Delphi method inputs and informed selection of issues and decision criteria. The context includes learning and critical thinking on the part of citizens or participants. The node in the right segment marks final multi-criteria decision-taking from inputs of four planning scenarios and four frames of thought. The context includes methods specialists and accountability reports as an output. Re-reflection repeats the entire process even after a planned or unplanned temporal break. Notice that the decision criteria derived from open epistemic democracy in the left segment create functional path dependence for the right segment. But both phases are subject to the vagaries of time. For example, shorter decision windows and election cycles favour motivational reasoning, whereas longer timeframes favour more deliberative thinking due to the informational tipping point effect.

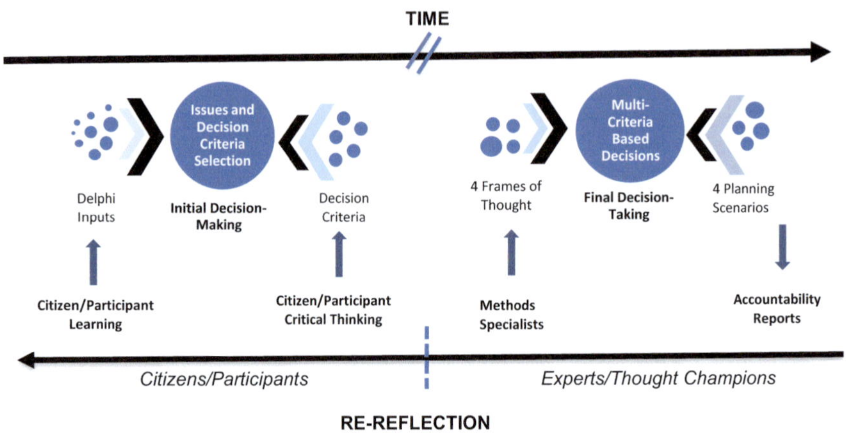

Fig. 10.2 The collective/democratic decision model

From a risk perspective, actions taken to mitigate risks must account for appropriate timing; otherwise, even the most deliberate decisions will fail. Achieving effective timing requires an understanding of when an outcome is needed and how long it will take to implement the response. More specifically, we must first understand how long the implementation of our action will take, and before that, we must determine how much time our decision-making process, including communication across channels requires [112, pp. 159–160].

In the international arena, the temporal break can be substantial, awaiting, for example, treaty ratification by legislative bodies. In South Africa, the commencement of legislation may be activated, in whole or in part, by a specified date or even left open-ended and up to the discretion of the relevant minister. The impacts of time between decision-making and decision-taking are negligibly researched in the literature to the point that what can be productively investigated as two distinct phases, each with its own unique characteristics, is mostly conflated in favour of decision-making. The closest discussion that I could find on the matter comes from *Our World in Data* which has a section on state capacity.[2] Their *state capacity index* captures the extent to which a state controls its territory, has skilled and impartial public servants, and is able to reliably raise sufficient resources. Ideally, however, a more coherent set of indicators is needed together with a time component that shows lag times between decision-making and decision-taking. A temporal break between decision-making and decision-taking can arise or be a part of the *design* of decision processes together with transparency provisions to counter possible negative group dynamics. For instance, group members can be given an "off-line" opportunity to submit comments and suggestions post initial decision-making but before final decision-taking. This can allow *citizen learning* and *citizen critical thinking* to better inform any subsequent requested inputs for planning scenarios, or other means

[2] https://ourworldindata.org/state-capacity

of democratic participation. In time, *after* final decision-taking, the probability of reaching a productive epistemic equilibrium between citizens and experts is likely to have increased, which may then be followed by implementation monitoring through *accountability reporting* per our model.

Re-reflections in the proposed schema can be sequenced in four bi-directional phases: decision-making then to decision-taking (left node to right node), or decision-taking back to decision-making (right node to left node), as well as exclusively in one node (left only) or the other (right only) in what can theoretically be an infinite number of iterations for non-urgent decision-making. Since solutions to problems will tend to be imperfect at first, they need to be sufficiently open-ended to benefit from lessons learned. Unlike mathematics problems, public issues are not amenable to binary solutions that are either correct or incorrect. Instead, they become more effective over time as new solutions are discovered or invented, and knowledge and skills cumulate. In the decision-making phase, deliberation can diminish group efficiency but improve final decision effectiveness by, among things, taking a wholeness perspective and better tracing the interrelatedness of important variables. For routine or recurring decisions, *Robert's Rules of Order* (or similar) can add efficiency to group deliberations but detract from the overall effectiveness of final decisions by, among other things, stifling design thinking and emergence. The four proposed frames of thought become even more important for non-routine decisions. In the decision-taking phase, there are many ways of ensuring that implementation is better infused by learning, such as careful monitoring and measurement, project and programme evaluations, implementation feedback from agents and beneficiaries, "quality circles", and other ways of encouraging the people involved at every level to think about improvement [24, p. 155]. It is important to note that the recommended decision-making and decision-taking methodologies in Fig. 10.2 are intended as prototypes for further development.

In the late 1990s, the concept of triple bottom line (TBL) reporting emerged with the aim of reporting on an organisation's economic, social, and environmental impacts. Since then, sustainability reporting has developed considerably due to the outbreak of COVID-19, with the *International Sustainability Standards Board* issuing its first two standards in 2023. Key to the viability of these and forthcoming standards is the concept of *integrated reporting*, which takes a stakeholder-inclusive approach, and also incorporates the "six capitals" of financial, manufactured, human, intellectual, natural, as well as social and relationship factors. Importantly, the newly established Board's objective is to establish a global baseline for sustainability reporting by issuing "exposure drafts" of their proposed standards [291]. This procedure, however, is likely to be too slow given the urgency required. Furthermore, its role as a "voluntary resource" and its lack of a definition for "sustainability", as noted earlier, may prove to be problematic in the foreseeable future but nevertheless ought to be vigorously pursued as a constituent part of a solutions package.

Decision Group Size

Two points of view, or a whole body of experts, are no match for one man of average ability exercising his independent judgement—Euripides

Elinor Ostrom has identified both group size and composition as two key variables that reliably influence the outcomes of collective decision-making. More specifically: the total number of decision-makers; the minimum number of participants necessary to achieve the collective benefit; similarities of interest; the presence of participants with substantial leadership or other assets; and the discount rate in use. Interestingly, she concludes that the same variables are also relevant to understanding the supply of institutions since they clearly represent a problem of collective action [126, p. 188]. Importantly, *the optimal size of decision groups for specific domains and tasks is currently unknown*, but their *generic composition* should follow the general lines of the "cognitive model" discussed earlier. The complexities arising from the participation of a large number of nation-states—more than 180 in the case of climate change—are mostly handled by the formation of a small number of coalitions that become the main protagonists in the process. Climate change negotiations have been dominated by four coalitions: the umbrella group; the European Union; the group of 77 plus China; and the Alliance of Small Island States. Institutional bargaining is generally a mixed-motive process featuring substantial parts of integrative and positional bargaining. Those seeking to arrive at mutually acceptable formulas under such conditions proceed by developing and refining texts that become the vehicles for moving negotiations from disparate perspectives to specific provisions that all major coalitions are willing to accept as components of a constitutional pact [37, p. 123].

Group size is a determinant factor in decision-making and decision-taking. Research with athletic teams, business groups, and even armies around the world has revealed that for a working group the ideal size is between eight and twelve individuals. This is natural because humans evolved as primates while living in small groups. Eight to twelve individuals can know each other well enough to maximise their capabilities and for organic leadership to develop. Beyond this group size, the possible combinations of communication between individuals get too complex and tasks need to been broken down. Commitment and participation fall off in larger groups, mobility suffers, and leadership tends to become manipulative and political [292, p. 203]. If participant's connections are too rigid, they may lose the incentive to cooperate if the connections cannot be loosened or broken. Similarly, if connections change frequently or are unstable, the tendency is to become less cooperative as noted earlier. Optimal cooperation in groups occurs when there is intermediate fluidity. The more connections participants have, the more they must be incentivised to cooperate. Research has shown that individual benefits must be twice the costs for group cooperation to arise because cooperation is harder in larger groups; hence, the payoff has to be of proportional worth [28, pp. 107–108]. But it is important to appreciate that this is only the inside-looking-out view at the scale from individuals to groups, what is still missing is the outside-looking-in view at the scale from groups to the collective. It is likely that the most promising solutions lie at this intersection. In the words of Geoff Mulgan: "how can the collective sing through the individual and vice versa?"

It would be particularly helpful to have a classification system by function that uses group size as one criterion of organisational design. Consider the following group size categories as a testable start: "small groups", 2–12 participants (e.g., BP corporate board); "medium groups", 13–25 participants (e.g., "Citizen Jury" participation method); "large groups", 26–100 participants (e.g., C40 group of cities); "small epistemic assemblies", 101–500 (e.g., IPCC Working Groups); "open medium-publics", 501–1000 (e.g., Iceland's National Forum); and "open large-publics", 1001 ~ (e.g., France's *Cap Collectif*). Wise collective decision-making may emanate from these groups if they are voluntary, collaborative, legitimate, authoritative, and transparent in their deliberations based on each member's implicit and explicit knowledge. And in the age of online meetings, not enough is known as to how group size may affect negative social influences compared to traditional in-person meetings.

Decision Group Composition

Dialogue is the collective way of opening up judgements and assumptions—David Bohm

Within organisations there are likely to be specialised groups that meet to make decisions formally. There is no set formula for such meetings, but experience suggests that there are strong correlations among the number of participants, their available knowledge and experience, the complexity of the task, the allotted time for decision-making, and the degree of a shared understanding and terminology for the task at hand. This is especially the case if a meeting is meant to agree on a resolution or make a decision. The most common reason organisation meetings fail is that they don't conform with meeting mathematics: there are too many people, or too little time, too little relevant knowledge or experience, too broad a topic, or insufficient common ground. Good meetings and decisions encourage the explicit articulation and interrogation of arguments and ideally allocate participants specific roles to probe them. Preferably these roles can be formalised, or may be left informal, as long as the uncomfortable aspects of disagreement are not minimised or overlooked [24, pp. 135–139], and signalling and reputational capital are used to good effect. Organisational rules also set important parameters for decision-making groups. For example, if the rules establishing a particular decision-making group call for final decisions to be arrived at by majority vote, then an odd number of voting members might work best. Some group size variables may also be mirrored to different degrees in the size of a required quorum, whether quorum size is set by an absolute number, or a specified proportion of the larger mandated group. Rules can also stipulate other essential procedural requirements: decisions can be prescribed to be binding if reached by consensus, majority, or supermajority prerequisites, while voluntary groupings may provide decision opt-out provisions to maintain cooperation and membership. "Terms of Reference" for the commissioning of external goods and services from office cleaning to nuclear procurement advisors extend the rules of an organisation to the outside world. Rules may also set out term limits and specify group composition criteria, including for sub-groups and any delegations of authority that they may require. Importantly, division of labour benefits can be gained by specialised roles assigned to particular sub-groups, such as a working group tasked with a specific objective, such as conducting sustainability research.

An organisational consideration for improving decision-making revolves around the criteria adopted for the formation of *fusion cells,* which are a type of sub-group designed to fuse intelligence across handfuls of personnel for a common objective in physical or virtual spaces. The cell acts as a network of teams to create synergies and to scale up needed capacities. Organisations can create fusion cells across organisational structures ensuring that people with different roles and knowledge are involved in key decision- making processes and the sharing of information, particularly by the provision of status reports to ensure timely action [112, pp. 116–117].

So far, our discussion on the composition of decision groups has largely inferred a top-down context for the selection or appointment of decision-group members with the minimum design requirement being a mix of diversity and expertise. Beyond these parameters, however, a bottom-up, self-selection option may be preferable for open public or democratic participation. One idea to enhance group composition diversity for public policy purposes is to utilise a small section of a jurisdiction's tax form to ask a taxpayer to indicate whether he or she would be willing to participate in a policy discussion forum of interest to her. Probably pertinent laws and prescripts will need to be amended and harmonised when it comes to implementation. A line item on the relevant tax form, for example, could look something like this:

> *Are you willing to participate in policy deliberations which are led by authorised organisations or persons that directly affect you: Yes __ or No__.*
>
> *If "yes", which one policy area would you be most likely to participate in (state in your own words):*

Taxpayers with privacy concerns can choose either not to answer these questions or accede to the stipulation that the information will only be used for the stated purpose. At a minimum, this would entail matching taxpayers with their expressed policy interests, and upon evaluating requests, this information could be shared with vetted entities that have a valid research purpose or other public interest objective. Resultant studies should then disclose the methodologies used, including any limitations such as the necessary exclusion of non-tax paying citizens. To ensure transparency, government custodians of such programmes could annually publish by policy domains and geographical areas the research groups that have accessed such authorised citizen participation lists. A more traditional route than using the tax system is to incorporate a suitable question in the household surveys that statistical agencies undertake in many jurisdictions.

De-Risking Decision-Making

Rationality is avoidance of systemic ruin —Nassim Taleb

We now have come from "evolution" and arrived at "constitution" by way of a possible referendum question that can be brought to the attention of citizens as follows:

Should ____ write or amend its constitution such that sustainability is included as a human right with subsidiary legal rights so as to perpetuate all life on Earth?

In answering and acting on this question, it should be kept in mind that existential global risk is ergodic. As Peter Railton has noted, "in the game of life, life must win every moment of every day, while death has to win only once". The concept of ergodicity is an important idea in many fields, such as physics, statistical mechanics, mathematics, statistics, economics, probability theory, finance, and decision-making, as well as to model the behaviour of complex systems. In an ergodic system the time-averaged activity of a single episode is equivalent to the ensemble-averaged activity of many episodes.[3] On a centennial or greater scale, the harm wrought by individual human violence for selfish motives is miniscule compared to the holocaust from self-transcending devotion to particular collectively shared belief systems. The resultant damage is derived from simplistic affiliation with these beliefs instead of mature social integration; it involves the partial surrender of personal responsibility and induces the quasi-hypnotic manifestations of group psychology [10, p. 266].

How people take risks, typically measured by gamble events in experiments, is central to our understanding of how we make decisions. In a 2021 study in PLOS Computational Biology by David Meder and his associates, the authors demonstrated quantitatively that the level of risk-taking changes with different dynamics as predicted from first principles of ergodic theory. Current theories of decision-making generally assume that preferences for risk are both idiosyncratic to individuals and stable over time. *Time optimality strategies*, however, go counter to existing expected utility theory, including prospect theory. The authors demonstrated how agents could compute averages when making decisions so as to *decouple* the ergodicity between the average of a system's possible states and its average over time such that *dynamical conditioning* can be learned passively and then *actively chosen*. Under multiplicative dynamics, the majority of decision-makers will have experienced negative time average growth at any one time, but a minority of decision-makers will have experienced extreme "wealth" growth such that it dominates their overall expectation value. In this gamble, however, maximising the expected value eventually leads to ruin for *all* players [293]. This is analogous to the concept of *mutual fate control* in social psychology whereby all members of a group affect one another's fate even if they cannot force others to behave in a certain way [28, p. 268]. According to Nicholas Taleb, there is a lack of appreciation of the

[3] Drawn in part from ChatGPT on 17.2.2023.

critical difference between conventional probability and path-dependent probability, which is a mistake that has persisted in economics and psychology since "age immemorial". He notes: "I am renewable, not humanity or the ecosystem. Ecocide, the irreversible destruction of our habitable planet is the one to worry about" [171, pp. 222, 228–229, 249]. And to do so it is essential to actively shift and explicitly communicate perspectives whether they entail discrete or continuous probabilities, objective-reflective-sustainability thinking, or as this book has tried to show, continuous mixed-scanning from the individual to the collective and back again.

To accommodate "dynamical conditioning" and implement "time optimality strategies" consistent with the cited PLOS Computational Biology study, it is also important to outline the most likely *informational contexts* in which these may occur. Table 10.1 represents a model state space onto which a timeframe can be superimposed thereby creating a risk reducing decision-making strategy by means of exploring as many states (cells) as possible with the view of combining them optimally. This would go towards satisfying the "dynamical conditioning" and "actively chosen" conditions of the PLOS study, which when simultaneously decoupled from the total time it would take to explore all of the possible model

Table 10.1 The information state space in sustainability decision-making

State of being	State of thought	State of knowledge	State of decision-making
Random Haphazard forces Such as gas particles in a container	*Emergence* Thought process element Dominant in new states, Determinant of opportunity	*Unknowable* The realm of randomness Causes are not discernible; effects are not predictable	*Sophia* The realm of higher wisdom Positive impacts of decision-taking arise
Complex Multiple forces Such as a city	*Relatedness* Thought structure element Dominant in dynamic states, determinant of fit	*Unknown* The realm of social systems Cause and effect may be determined after events	*Phronesis* The realm of practical wisdom Effectual decision-taking among options
Periodic Regular forces Such as earth's day and night cycles	*Wholeness* Thought structure element Dominant in episodic states, determinant of parameters	*Knowable* The realm of scientific inquiry Cause and effect can be determined with sufficient data	*Episteme* The realm of formal verification Institutionalised learning and supported decision-making
Equilibrium Balance of forces Such as an ecological predator-prey balance	*Design* Thought process element Dominant in stable states, Determinant of frames	*Known* The realm of scientific knowledge Cause and effect are understood and predictable	*Techne* The realm of codified applications Rule-bound learning and decision-making

Sources: "State of Being", [294]; "State of Thought"; Author; "State of Knowledge", [240]; and "State of Decision-Making", Aristotle states can potentially lead to the best solution within the available time and information states.

In addition to broad ergodic strategies, experts have proffered operational advice as shown in Box 10.1 to help remedy negative decision tendencies. But perhaps more instructive is India's Prime Minister Narendra Modi's description of his own decision-making process reproduced in Appendix III. Beyond such inspiration and insights, it is also clear that many areas of collective decision-making relevant to sustainability issues require a new and sustained research agenda, including the application of an ergodicity frame and *directed minmax attention* up to and including catastrophic climate change scenarios [253]. Such research strategies, however, will take up time that we do not have. Time is the resource in shortest supply—consequently the research agenda must focus on rapid solutions [98].

Box 10.1 De-risking Group Decision-Making—Advice from Experts

The mediating assessment protocol [82, p. 323].

The mediating assessment protocol is a decision-making technique that can be applied broadly whenever the evaluation of an option requires considering multiple dimensions. Mediating assessments should be assigned to different team members to produce about seven mediating assessment compilations covering the most important variables that affect the decision. A second team should then be tasked with the independent evaluation of these assessments and reach a conclusion in the form of a rating (steps 1–3). In the second phase of the process (steps 4–6), the "estimate-talk-estimate" method can be used to elicit the average of independent opinions together with systematic deliberation, including paying attention to any potential "deal breakers" or "deal clinchers". More specifically, the following steps should be followed:

1. At the beginning of the process, structure the decision into discrete preliminary assessments.
2. Ensure that whenever possible, these mediating assessments use an outside view.
3. In the analytical phase, keep the assessments as independent of one another as possible.
4. In the decision-making meeting, review each assessment separately.
5. On each assessment, ensure that participants make their judgements individually; then use the "estimate-talk-estimate" method.
6. To make the final decision, delay intuition, but don't ban it.

Combatting "group think" [234, pp. 105, 134]

To improve group decision-making, further to points 1–6, the most significant criteria for an optimal solution should be canvassed at the identification stage and its subsequent review (similar to steps 1–3 above but more granular), including *consultation with stakeholders or citizen beneficiaries* (an important additional element). Manage bias by controlling social factors such as status, talkativeness, and the likeability of persons providing feedback. *Appendix II is a tool to combat biases in real time.* Also consider an anonymous procedure, such as secret balloting or the Delphi technique. Adopt a credible method of aggregating individual independent evaluations into an acceptable consensus, such as in steps 4–5. Additionally:

7. Select and empower the inquisitive yet self-silencing leader.
8. Prime critical thinking in the group.
9. Provide early feedback regarding group successes.
10. Assign roles to designated champions.
11. Adopt multiple perspectives.
12. Allow "devil's advocates".
13. Incorporate the Delphi group of decision-making methods.
14. Consider employing "red teams".

De-Risking Decision-Taking

All persuasion is self-persuasion—David McRaney

In scaling-up de-risking from the decision group to the main collective, shared intentionality for sustainability and open epistemic democracy can be brought back into focus again. More specifically, evolutionary "group-mindedness", properly fashioned, can lead to collective epistemic conditions in the service of the "open democratic" part of the Collective/Democratic Decision Model. Stated differently, participant diversity and requisite expertise, subject to the crucible of democracy, can help de-risk collective decision-taking. The heyday that treated international society as a system of sovereign territories has been relatively short. However, this does not imply that the existing society of states, with its elaborate protections for the rights of individual nations, is likely to collapse and be superseded by some alternative form of social organisation at the global level. What is more likely is the emergence of more complex arrangements in which new actors and new social practices become important to the development of strategies designed to address global challenges. International society lacks a mechanism or public authority of the type required to establish and administer service organisations dealing with matters such as compliance and adjudication of claims, which places a significant burden on international regimes with such mandates relative to their domestic counterparts. What is needed is a targeted effort to design and implement institutional arrangements that take local knowledge seriously and protect the rights and interests of local stakeholders, even while they introduce mechanisms at higher levels of social organisation required to address the dynamics of ecosystems up to and including the global level [37, pp. 107, 127, 144, 158]. One such possible effort can be the creation of an analogous international version of the Overton Window of Political Possibility to provide information on the ratio of opponents to proponents regarding sustainability issues in representative stakeholder groups, as well as the proportion of epistemic to diversity indicators in the same stakeholder groups.

The new SURFPAD (Substantial Uncertainty with Ramified or Forked Priors or Assumptions = Disagreement) framework developed by Pascal Wallisch and his associates at New York University is emerging as an alternative explanatory model of group decision-making. Starting at the level of perception, different life experiences can lead to very different disambiguations of ambiguous information given that participants are not aware of their own, nor others' experiences and brain processes so as to understand and arrive at an agreed to conclusion. In psychology this is known as naïve realism which leads many of us to believe that we arrived at our beliefs, attitudes, and values after careful analysis and not through unmediated thoughts and perceptions, only aware of the result and not the processes. As David Bohm has noted "thought moves towards deception". According to Pascal Wallisch "we need to be able to say 'I am subject to SURFPAD, you are subject to SURFPAD'. If we can do that, we go to a meta level". On this path we can focus attention on the *how* and the *why*, and not the *what*. It is literally how minds change [179, pp. 50, 72–74, 80, 86–87]. More concisely, different assumptions at the subconscious level create path dependence at the conscious level subject to the interpretation of evidence by the interplay of Bayesian priors and Bayesian posteriors.

Even if wrong, every individual perspective has value when it is a part of a decision-making or decision-taking group that can interact and communicate. Humans are social animals first, and individual reasoners second, a biological system built by evolution and then expanded via a psychological system of reasoning that grew under selective pressures to facilitate misinformation. Under such conditions, confirmation bias turns out to be prevalent and even useful. Reasoning is biased in favour of the reason owner, and that is important because each person needs to contribute a strongly biased perspective to the pool of perspectives since the effort can be off-loaded to the division of cognitive labour by the group. A participant is well served even by quick decisions and intuition since these will be subject to group evaluation. If others produce credible counterarguments, thinking can be refined and priors updated. Members of the group who see the correct path from moment to moment can guide others and in turn be guided by them. Research has found that *reputation management* is what creates group cohesion to the extent that if accepting certain facts could damage personal reputation, strong resistance to updating priors sets in. However, the threat to one's reputation can be ameliorated by revisiting deep personal values, or affirming a separate group identity. In the case of scientists, doctors, and academics, a virtuous circle exists because both their identity and belonging goals are met by pursuing the accuracy goals of their respective professions. Their openness to change and willingness to question orthodoxy, as well as the beliefs of others, signal their fidelity to their particular group. By some accounts, however, we live in a post-*trust* world where a general mistrust of media, science, medicine, and government prevails and where objective information is trumped because certain groups *trust* only sources that share *their* views—this is convergent thinking without the benefit of divergent thinking. This bifurcation of thinking becomes bifurcation of communication as well. Psychologists Philipp Schmid and Cornelia Betsch have characterised persuasive communication as being one of two types: "topic rebuttal" and "technique rebuttal". Topic rebuttal is the preferred method in trusting environments where the evidentiary standards have already been established and in-group members rely on facts around a topic, say, in a technology laboratory. Conversely, technique rebuttal involves asking people to step backward through their own processing to understand how they arrived at a conclusion and whether their reasoning was sound, say, about student loan forgiveness. Technique rebuttal works due to the phenomenon of an affective "tipping point". According to David Redlawsk and his team, this is the moment in which people can no longer justify ignoring or discounting a stream of disconfirmatory evidence. Once a person reaches his or her affective tipping point, the brain switches from conservation to active learning. One early indication of where the tipping point might lie is when 30% of the incoming information is incongruent with one's stored information. In practice, many other variables may be activated placing the threshold at a different level, or making it unreachable. The important take away is not the actual number, but the existence of the threshold itself, implying that minds can be changed with a credible amount of evidence. An important caveat, however, is that incongruences make us feel more certain, not less *before* an affective tipping point is reached [179, pp. 119–227].

One method of technique rebuttal is "street epistemology", which is suitable for processing both objective and subjective claims, including issues of sustainability. According to Anthony Magnabosco who developed the method, the key is guided metacognition through the encouragement of a person to think about their own thinking, but only after they have used their own reasoning to arrive at a claim together with its justification. With facilitation, they can then evaluate their own methods, question their own reasons, and assess the merits of their own argument. The process involves nine steps: (1) establish rapport; (2) ask for a claim; (3) confirm the claim by repeating it back in your own words; (4) clarify their definitions and use only them; (4) ask for a numerical measure of confidence in their claim; (5) ask what reasons they have for holding that level of confidence; (6) ask by what method they quality assured their own reasons; (7) use the same method to (8) listen, summarise, and confirm; and (9) synthesise and conclude on a positive note [179, pp. 226–227]. The Street Epistemology website has more detailed information.[4] It would also be helpful to combat unwarranted *cognitive ease.* When one feels strained by required effort and unmet needs they are more likely to be vigilant and suspicious, invest more effort on the task at hand, feel less comfortable, and make fewer errors, but also be less intuitive and less creative than normal. Concomitant ease arises from primed ideas, previous experience, a good state of mind, and the use of clear language and display [85, pp. 59–60]. Even a clear, bold font can promote cognitive ease.

[4] https://streetepistemology.com/resources

By way of review, first consider that technique rebuttal (the more challenging option than topic rebuttal) would likely be more effective on a one-to-one basis instead of a group setting where it might be perceived as a type of naming and shaming. Second, it would take much longer to explore the attitudes of several individuals as opposed to one and probably become unfeasible for larger groups. More research is needed to establish whether group tipping points exist (say, upon disconfirming information of 30%) and under what conditions they might be utilised fruitfully. Third, as the epigraph of this section informs us, "all persuasion is self-persuasion", which infers that the most useful skill is probably assuring the listener(s) that you, as the facilitator, are open to having *your* mind changed. We can all demonstrate openness by acknowledging when we agree with our critics and by giving them credit for any lessons learned. If we then ask what views they might be willing to change, we are not likely to be perceived as hypocrites [19, p. 107]. Finally, other issues such as privacy, lost trust, deadlines, and so on would also come into play, making a wholeness strategy particularly important if minds are to be changed.

In Box 10.2, experts have proposed a number of operational factors for groups to consider before final decision-taking.

Box 10.2 De-risking Group Decision-taking—Advice from Experts*

1. *Sound decisions are timely.* Your assessment of risk will dictate whether you should take a more or less cautious approach. Arriving at the right balance between short-term and long-term considerations is key to unlocking true value. Although the decision-maker must urgently address symptoms, once this is done, a plan to fix the root cause should be developed.
2. *Do not shy away from contingencies and caveats.* Advocate for reasonable precaution, not unreasonable caution. Use personal judgement and a cost-benefit analysis to chart the best course forward with the information received. Acknowledging competing claims and conflicting results does not diminish one's credibility. Sound decisions are made by a clearly accountable person.
3. *"Complexify" contentious topics.* Look at them as if through the many lenses of a prism because seeing "shades of grey" makes the group more open minded. Consult with those who can contribute the most in a meaningful way. Expand your emotional range by exhibiting more curiosity, or even admitting confusion or ambivalence. You're looking to bring everyone up to a consistent level of understanding, which is often necessary for the smooth implementation of a major decision.
4. *Keep a "rethinking scorecard"*. Establish decision-taking (evaluation) committees of about 5 members. Do not evaluate decisions based only on the results, also track how thoroughly different options were considered. Sound decisions consider the holistic impacts of a problem and are shaped

(continued)

Box 10.2 (continued)

by consideration of many different viewpoints. Sound decisions are made as close as possible to the action.

5. *Rethink actions, not just inputs*. Do not unnecessarily refer back to matters already decided or insist on doing everything through formal channels. If appropriate, avoid calls for "further study and consideration".
6. *Sound decisions are communicated well to stakeholders***.** Talk economically, on-point, and minimise personal experiences. Communicating the substance and reasoning to stakeholders behind the decisions taken is critical for obtaining their buy-in. Avoid "wordsmithing" communications, minutes, and resolutions.

Adapted from [19, pp. 251–257], [295] and the former U.S. Office of Strategic Services

Chapter 11
Epilogue and Way Forward

As our thoughts are, so will our lives be—Seneca

According to Ian McGilchrist's thesis, the Western World is becoming increasingly left-brain dominated, prompting him to presage many societal conditions that currently prevail, more so in some places more than in others. He predicted that ever more narrowly focused attention on parts rather than the whole would lead to greater specialisation and technicalising of knowledge. Skilled roles within society would become the object of suspicion and derision. Skills themselves would be reduced to algorithmic procedures, which could be drawn up, and regulated if necessary, by bureaucrats. Tacit forms of knowledge would become dispensable accompanied by a rise of explicitness supported by ever increasing legislation. Knowledge would be replaced by information and its gathering. Knowledge itself would seem more real than what we might call wisdom, which would seem too nebulous, something never to be attained. There would be a focus on material things as opposed to the living. Nature not managed or subject to exploitation would be seen as a threat requiring bureaucratic control. And we would expect there to be a resentment of and deliberate undermining of our sense of awe and wonder [5, pp. 428-434].

Today's world elites, having understood the climate crisis, but realising that they would have to pay dearly in the future, have chosen instead to make any who are not equally enlightened pay for it now by denying its very existence. As imaginatively told by Bruno Lamour in his essay *Down to Earth* and exemplified by Trumpian politics (which is not post-truth but rather post-trust since it rejects the political world that it claims to inhabit), what would we know of the so-called New Climate Regime, especially since it has no institutional embodiment? The elites in the New Climactic Regime have to learn to cohabit the Earth with those whom they used to deem archaic, traditionalists, or reactionaries [75]. *Sustainability Thinking*, if not a precise antithesis of this rather bleak picture, is nonetheless an attempt at forestalling or preventing variously evident self-harming trajectories, particularly in the West and developed North. By way of contrast, in Japanese culture, the word, *jibun,* implies a share of something which is both separate and not separate, individual and yet still shared. Everything about the Japanese attitude towards nature, expressed both in their mythology and in everyday life, suggests a stance of mutual trust, dependence and interrelationship between man and nature—*shizen*. A reverent

E. Bellos, *Sustainability Thinking*, https://doi.org/10.1007/978-3-032-11244-6_11

attitude towards *shizen*, now absent in the West, is characteristic of even the Japanese scientific education system [5, pp. 453, 456].

Often political opportunities to create momentum for system level transformation arise out of crises. Both a perceived threat and the real emergence of a crisis together with any associated public remonstrations can serve as a window of opportunity to use systems approaches to reconfigure public service organisations and policy at the highest level. Emergencies tend to suspend the norms and rules that limit experimentation. Most importantly, a crisis can be an opportunity to revisit the central objectives of policies, programmes, and services. By questioning and reasserting purpose, a public administration creates the prospect of redesigning not only services, but how those services are resourced, managed, and reinstituted if and when the crisis recedes or ends [96, p. 29].

From Sustainability Thinking to Sustainable Action

Because action always stretches forward in time, so must a mind that reliably succeeds in action—Peter Railton

Prospection is the actual ability that makes the aspiration of wisdom a possibility [251, p. x]. It has come about because of our unprecedented capacity for anticipatory guidance together with our prodigious capacity to live and learn from others. In short, we learn to be wise [9, p. 7]. According to Martin Seligman and his co-authors, the proclivity and talent of humans for prospection is a fundamental human characteristic. Prospection consists of two steps that can be summed up as: (1) what do I want? and (2) what can go wrong? Thinking about the future is about getting what you want, so when one imagines the future, the tendency is to view it in a positive light, but upon further reflection it becomes evident that plenty of things can go wrong. The future is uncertain, the path chosen to reach one's goals is fraught with threats and risks that could stop progress and the attainment of desired outcomes [296, p. 176]. The first step of prospection is underpinned by *optimism bias*, while the second step predisposes one to *loss aversion*—and both are subject to inherent tension one with the other. These competing psychological tendencies at the individual level and their manifestation and cross-play at the group level are basic to decision-making and central to this book's thesis.

Our brains evolved to survive more so than to think, which explains *biophilia*—the innate love of living things and humanity's affinity with our ancestral environment [297, pp. 202, 211]. In consequence, sustainability thinking should focus on providing people with opportunities for action to allow them to start a self-persuasive process so as to develop agency together with others. Rather than telling them what to do, sustainability information should support them in discovering *how* to act. Communication campaigns which focus on building awareness or concern do not lead to the same deep engagement as self-persuasion, nor set people on such a path. Counter to conventional wisdom, renewed social psychology research suggests that our actions can change our beliefs rather than the other way around. Action on action leads to self-persuasion by learning from the actions of others. Research on bystanders shows that protest marches increase positive beliefs about the marchers and collective efficacy. A shift of mind from issue-based to action-based activity is needed to create for citizens the opportunities and capabilities to engage in action which is experienced as meaningful. The top-down approach from behavioural economics popularised by books like *Nudge* by Richard Thaler and Cass Sunstein minimises personal agency and hence the meaning and reward derived from that agency, including the need to become more self-informed and self-reliant. For instance, top-down imposition of climate policies by governments can translate into local opposition if perceived to be unjust or lacking in transparency. Power and agency shape which sustainability narratives are told and how prevalent they are. Indigenous communities, for example, have used their own narratives to imagine climate futures divergent from top-down government-led narratives. Moreover, paying attention to the importance of informal economies opens up many possibilities for new approaches to service provision and decent living standards along with climate resilience. When an individual's pro-environmental behaviour is associated with personal meaning, then it also increases subjective well-being, especially if coupled with feedback as to whether one's own actions compare favourably with what other people do, think, or expect. Consider that a rooftop solar installation spawns copycat installations in the same neighbourhood within 4 months, on average [298, 299].

It is easy to become overwhelmed and surrender trust to political representatives to make the right decisions on our behalf yet they themselves might need the help that we can provide. One tried and true tool is a *model law* which any citizen or group of citizens can write. A model law is draft legislation in the form of a completed template that can be used to encapsulate the views and positions of groups and organisations for the further attention of political representatives. The Finish Law on Citizens' Initiatives reproduced in Appendix I is an excellent example because it provides for both the process and outcome of effectively targeting and furnishing such inputs. Writing itself is the underlying mechanism by which to organise thinking and thoughtful action. The process of writing confers a cognitive bonus as opposed to simply watching or hearing, which impose few distinctions on the streaming stimuli. When we begin writing, however, we are forced to differentiate, judge, and select. This more engaged mental activity leads us to process what we are paying attention to more deeply and can also lead us to have *new thoughts* [13, p. 152].

Accordingly, citizen philosophy, as envisioned here, is a call to ordinary citizens to be publicly engaged as much as possible beyond their personal spheres so as to become smarter than their respective leaders as a means of making them listen more attentively, patiently, and humbly about our common future. Key decisions should not be made just by politicians and scientists; they matter to all of us and should be the outcome of effective public participation and debate. The needed enabling conditions are enough "feel" for the key ideas of science and enough numeracy to assess hazards, risks, and probabilities so as not to be fooled by experts, or be taken in by populist mantras [300]. More specifically, sustainable action can entail any combination of the items listed in Box 11.1.

Box 11.1 A "To-Do-List" for Sustainability and Self-help Democracy

- **Vote** in local, national, and regional elections if and when voting is made available. Assume and act as if you are required to do so by law (in several countries like Australia, Brazil, and Cyprus voting is compulsory).
- **Run for public office** if and when available**.** Adopt a bottom-up strategy based on your track record at smaller jurisdictions. Find and share the information about how each sphere of representation works in your jurisdictions.
- **Form or join a community group** and provide timely community relevant inputs to them and through them.
- **Participate** in surveys, censuses, town halls, and other organised attempts to solicit your inputs (e.g., Delphi). **Listen attentively and speak-up in important public meetings**.
- **Provide thoughtful inputs** when called upon to do so by legislative or regulatory bodies in accordance with their stipulated deadlines and procedures. Follow-up with your legislators/administrators as a means of monitoring implementation.
- **Draft model laws**, or their equivalent, together with others as inputs to legislators. **Concentrate on cross-cutting issues** like electoral laws, freedom of information laws, decision-making criteria**,** public participation laws and regulations, environmental laws and regulations, consumer laws and regulations, energy efficiency standards, food and other labelling, and educational incentives.
- **Initiate a referendum or petition drive** concerning your most important issues in accordance with any relevant legal prescripts (see change.org, avaaz.org, and explore the D-CENT tools on the Internet).
- **Initiate a citizen planning scenario project** to capture your community's vision for the future and submit it as part of your inputs to legislators/administrators.
- **Report corruption and maladministration** when present via official channels if available.
- **Stay engaged with news affecting you and your community, especially "accountability/citizen reports"**, or their equivalents. **Ask your political representative to begin such an initiative** if none exists.
- **Start a non-profit organisation** or socially responsible business in accordance with your jurisdiction's requirements and your subjects of interest. Solicit financial and other support for doing so if need be.
- **Subscribe to research outlets and newsletters** to stay abreast of issues that interest you.

- **Start an honesty library,** or **human library,** or animal library, or plant library (or all) in your neighbourhood, and support your local libraries.
- **Become proficient at something you value and become a good citizen model** for your family, friends, colleagues, and others **regarding this "something".**
- **Participate in citizen science projects, ecotourism, "smart destination" tourism,** and **socially minded gaming,** if and when you can.
- **Consume less**, and if not, undertake any combination of the following if you can: buy "local", buy "green", and buy "ethical".
- **Donate money/goods/services/time** to your favourite charity/library/religious institution.
- **Consider donating your body/organs,** if permitted by your religious beliefs.
- **Agitate in the Socratic sense** (become visible and audible) where you live and for what you want ("organise, organise, organise") around these issues.
- **Learn-critique-create** (draw back critically to leap forward creatively) in all things.
- **Create your own "to-do-list".**

Additionally, *Project Drawdown* calls attention to 93 technologies and 20 high-impact actions that individuals can take to help with climate change, as well as a comprehensive list of references detailing various solutions.[1] Such initiatives, however, should ideally be "de-risked" with respect to the interplay between the "rebound effect" and "diffuse responsibility". In particular, the free-rider problem on the part of fossil fuel companies and others at the expense of the general public should be mitigated or averted by timeous and appropriate incentives or sanctions. For new sustainability policies and climate mitigation technologies to be rapidly and extensively implemented, they must be socially acceptable to those who are directly impacted by them. Participatory mechanisms that enable deliberation by a representative sample of the public can inform policymaking and increase the legitimacy of new and difficult policy decisions and actions [299].

Back to the Future from the End of History

Nine-tenths of wisdom consists in being wise in time—Theodore Roosevelt

According to UNESCO, there has been a growing demand for extramural philosophy, struggling for an identity, yet corresponding to a fundamental need in

[1] https://drawdown.org/references

society for something beyond institutionalised philosophy in universities and other enclaves. This revived need to philosophise, to think critically and creatively, has risen from the diminution of several traditional ways of life, whether in terms of politics, ideologies, morals, or religion [301, pp. 153–195]. Society is based on shared meanings; these shared meanings are the glue that holds society together, and one might say that our present society has some very poor-quality glue. And meanings are not static, they flow and in so doing they are shared becoming a vital glue, it is in this way that we can talk together coherently and think together [43, pp. 32–33,46]. Human societies, unlike animal societies, are based on the genetic propensity to form long-term contracts that evolve by culture into moral precepts, and finally, become law [23, p. 332]. As the late eminent biologist E. O. Wilson has observed presciently "philosophy plays a vital role in intellectual synthesis, and it keeps us alive to the power and continuity of thought through the centuries". We are learning that it is all about ethics. Natural selection requires us to integrate ethics and morality into science and technology and to consider their implications for human welfare, failing which humans may well go extinct. Once humans developed the capacity to over-exploit the environment, ethics became a necessity for survival [213, pp. 240, 242]. Today the ethics of scientific research is a major area of concern to many scientists. Climate change with its growing implications for disparate economic gain and international conflict requires more critical reflections by scientists about how their results can be manipulated and packaged by short-sighted politicians, multinational corporations, and nation-states for their own benefit but to the detriment of other stakeholders. The scientific strategy for an incisive understanding of the link between societies and the environment can consequently profit not only from an understanding of how a managing elite make decisions on environmental issues, but also from a multilevel approach that views the world in terms of interrelated and interactive cultural landscapes with human agents differentiated by geography and occupation yet unified by the urgent need to cooperate for their mutual benefit. At the global scale, an ideology of environmental and social ethics is needed that valorises intergenerational resilience and renewal of natural resources while concurrently enhancing the opportunities for a better life of those that have been historically disadvantaged [302, p. 189].

It is both a legal and philosophical question as to whether people ought to assert for themselves the principle of Natural Law as a first principle, but deny the same foundational right to nature herself. Witness the burgeoning rights of nature movement is pushing local, national, and international authorities to recognise natural features in law, giving them either legal personhood or an independent right to flourish. Giving rivers the status of people in courts of law is enlivening environmentalism around the world. Ecuador started the movement in 2008 when it enshrined rights of nature in its constitution. Bolivia, Mexico, and Colombia have created comparable legal mechanisms to protect nature, while New Zealand, Australia, and Bangladesh have acted to protect rivers. In the United States, residents of Toledo, Ohio, drew up a bill of rights for Lake Erie. Rights of nature are most prominently asserted in post-colonial countries where indigenous people strive to protect their traditional lands. But can legal standing for nature protect it in reality? The UN

Special Rapporteur on Human Rights and Environment is of the opinion that legal personhood could succeed where decades of environment laws have failed. Mumta Ito, founder of *Nature's Rights,* argues that ultimately the law must recognise that nature's rights come first, followed by human rights, and then corporate rights [303]. In 2022 the UN General Assembly adopted a historic resolution, declaring access to a clean, healthy, and sustainable environment, a universal human right. In doing so the UN Secretary-General noted that the decision will help States accelerate the implementation of their environmental and human rights obligations and commitments. Similarly, the UN High Commissioner for Human Rights added that environmental action based on human rights obligations provides vital guardrails for economic policies and underpins legal obligations to act, rather than simply relying on discretionary policy. The 2022 decision recognises that environmental damage has negative implications, both direct and indirect, for the effective enjoyment of all human rights, including the implication that the resolution will change the very nature of international human rights law [304]. Against this background, the International Court of Justice noted in 2025 that human rights have been recognised to be indivisible, interdependent, and interrelated and that the effective enjoyment of a number of human rights cannot be fully realised if those who hold them are unable to live in a clean, healthy, and sustainable environment [104, p. 112]. Accordingly, a new philosophy of thought is needed for our age that will take into account our evolutionary past together with our sustainable future because almost everything around us has been and is mediated by values and thought. Put differently: *a reverent philosophy is needed for new sustainability thinking*—we need to refocus our attention and explicitly acknowledge life itself as a preeminent value in its own right. It follows that life itself ought to be a *first* in collective human thinking and thought, and a *universal moral right* of all entities that possess it. This then implies that sustainability ought to be an *enforceable human right* as the prerequisite condition for the exercise of all other human rights. To realistically give legal effect to sustainability as a human right, it ought to be incumbent on nation-states to urgently incorporate relevant settled provisions in their national constitutions while the UN may in tandem initiate the process of establishing a UN Specialty Court for the purpose of equitable enforcement and compliance.

In conclusion, the guiding frameworks of the first law of democracy and open epistemic democracy as its preferred implementation mechanism together with the call to make sustainability a human right were advanced to help us move urgently, individually, and collectively, towards practical philosophy. To this end, citizen philosophy and thoughtful action were operationalised via a synthesised decision-making model centred on citizen learning, critical thinking, and wise decision-taking. In the proposed operational schema, the main practical challenge of each expert thought champion would be to devise ways and means to translate what might work in a one-to-one mode (champion to participant) up to a one-to-all mode (champion to participants of a stakeholder group and possibly up to all global citizens). In brief, an invitation to present and future citizen philosophers and thought champions has been tendered to build on (or refute) some or all of the book's central propositions so as to arrive at the essence of *phronesis*, or to establish new directions of inquiry, including contemplating the counterfactuals of democracy without sustainability, or life without love, or beauty.

Sustainability Thinking sought to encourage all global citizens to bring its title to life in whole and in part, in theory and in practice. In essence it attempted to learn wisdom by reflecting on real-world dilemmas that call for it [235, p. xiii]. It sought to elucidate the basic social conditions for wisdom-making and its decision-taking. From strategies of thinking to possible desirable futures, it is hoped that at least the aspirations and choices of "*Homo prospectus*" were expanded towards thoughtful action and freedom from lies. Whether, like the mythical Phoenix rising from the ashes, it managed to resurrect reverence for all life in the context of sustainability thinking and sustainable social action in an ideal epistemic democracy, it is a testament that both science and philosophy can inspire, but not separately guide the human agency and wisdom needed for human self-preservation, well-being, and environmental stewardship in the Anthropocene. Society will be well served by embracing multi-perspective valuations of life such that adaptability is effectively invoked to escape negative path dependency in science, economics, and politics, as well as to avert powerful private interests—including those stemming from AI technologists—from overwhelming global public welfare. Scientists should become more like ordinary citizens. They should act with confident humility and not rely merely on dispassionate facts. The time has come for scientists to organise and actively lobby policymakers with regard to their significant findings. On the other hand, citizens should become more like politicians, and politicians should become more like philosophers. I am a *citizen-philosopher*—you can be one too in the ecosystem of collective wisdom and thoughtful individual action that constitutes our common mindscape.

Appendices

Appendix I

Finnish Model Law on Citizens' Initiatives

Done at Helsinki, 13 January 2012

Citizens' Initiative Act (also as a Model Act)

In accordance with the decision of Parliament, the following is provided:

§ 1

Purpose and scope of the law

This law lays down the procedure to be followed for the adoption of a citizens' initiative provided for in the Constitution.

The launch of a European citizens 'initiative is governed by Regulation (EU) No 211/2011 of the European Parliament and of the Council on the citizens' initiative. Articles 10 and 15 of this Law shall also apply to the taking of a Union citizens' initiative.

§ 2

Right of initiative

The right to submit a citizens' initiative to Parliament is provided for in the Constitution.

§ 3

Initiator and representative of a citizens' initiative

The initiator of a citizens' initiative may be one or more Finnish citizens entitled to vote.

The initiator of the initiative shall nominate at least one representative and an alternate.

The initiator of the initiative and the contact information of the representative and deputy representative must be notified in connection with the transaction to the

E. Bellos, *Sustainability Thinking*, https://doi.org/10.1007/978-3-032-11244-6

Population Register Center, the Ministry of Justice, and Parliament for communication on the initiative.

§ 4

Form of the citizens' initiative

The citizens' initiative includes a bill or a proposal to start drafting a law as well as the reasons for the proposal. The initiative must have a title describing its content. The initiative must be dated.

Different issues must not be included in the same initiative.

The initiative is written in Finnish or Swedish, or both.

§ 5

Collection of statements of support

Signatures in support of a citizens' initiative (statements of support) are collected in paper form or electronically on a computer network. The form to be used for paper statements of support is regulated by a decree of the Ministry of Justice.

Statements of support must be collected within 6 months of the date of the initiative.

When collecting statements of support, the initiative, information on its initiator and representative, and the contact details of the representative must be available to the signatories of the initiative. The initiative may not be changed after the start of the collection of statements of support.

§ 6

Content of the statement of support

The statement of support must include:

1. The title and date of the citizens' initiative.
2. The full name, date of birth, and domicile of the signatory of the initiative.
3. A declaration by the signatory of the initiative that he or she is a Finnish citizen with the right to vote and that he or she has made a statement of support for the same initiative only once.
4. The date of the statement of support.

§ 7

Technical requirements for the electronic procedure

When collecting statements of support electronically in the information network, an information system approved by FICORA must be used. Approval is conditional on:

1. Strong electronic identification referred to in the Act on Strong Electronic Identification and Electronic Signatures (617/2009) is used in the collection of statements of support.
2. Unauthorized alteration and other unauthorized or inappropriate processing of the information contained in the system is prevented by appropriate and sufficient security arrangements and other measures for access control, access control and information networks, information systems and information services.
3. The system complies with the regulations issued by the Population Register Center referred to in Subsection 2 below.

The Population Register Center shall issue regulations on the standard format of the information collected in the information system, in which it can be electronically combined with the information in the population information system for the inspection provided for in Section 9.

There is no fee for approving the information system.

Information on FICORA's approval must be available to the signatories of a citizens' initiative on the website where statements of support are collected.

§ 8

Ministry of Justice online service

Statements of support can also be collected in the online service maintained by the Ministry of Justice. This is conditional on the citizens' initiative having already received at least fifty statements of support within 1 month of the date of the initiative and that the initiative meets the requirements laid down in this Act.

The online service must meet the technical requirements provided in Section 7.

There is no charge for using the online service.

§ 9

Verification of statements of support

The representative submits the statements of support to the Population Register Center, which checks their correctness and eligibility.

At the request of the representative, the Ministry of Justice submits the statements of support collected in the online service maintained by the Population Register Center for inspection.

The Population Register Center confirms the number of approved statements of support.

There is no charge for checking statements of support and confirming the number of statements of support accepted.

§ 10

Duties of public authorities in connection with the European Citizens' Initiative

The competent authority in Finland referred to in Article 15 (1) of Regulation (EU) No 211/2011 of the European Parliament and of the Council on the citizens' initiative is FICORA. The competent authority in Finland referred to in paragraph 2 of that Article is the Population Register Center.

§ 11

Submission of a citizens' initiative to Parliament

The representative shall forward the citizens' initiative and the decision of the Population Register Center issued on the basis of Section 9 to Parliament if the conditions for the right of initiative provided for in the Constitution and this Act are met.

The initiative shall lapse if it has not been submitted to Parliament no later than 6 months after the decision of the Population Register Center issued pursuant to Section 9.

§ 12

Personal information

Personal data collected in statements of support may only be used in the manner provided for in this Act. The initiator and his or her representative may disclose information only to the Population Register Center. The information held by the authorities will be public once the minimum number of statements of support laid down in the Constitution has been reached.

The initiator and representative of the initiative shall take care of the destruction of the statements of support in their possession no later than 6 months after the decision of the Population Register Center issued pursuant to Section 9. If statements of support are not submitted to the Population Register Center for inspection, they must be destroyed no later than 6 months after the collection of statements of support has ceased.

The Population Register Center must destroy the statements of support at the end of the parliamentary consideration of the initiative. If the initiative is not submitted to Parliament, the Population Register Center will keep statements of support for 9 months of the decision it has made pursuant to Section 9, after which they must be destroyed.

The Personal Data Act (523/1999) otherwise applies to the processing of personal data collected in statements of support.

§ 13

Notification of financial support

The initiator and the representative of a citizens' initiative shall ensure that the financial support received for the launch of the initiative and its donor are indicated if the value of an individual grant or a grant consisting of several services received from the same sponsor is at least EUR 1,500. The name of an individual may not be disclosed without his or her consent if the value of the support provided by him or her is less than EUR 1500.

Services received in cash, goods, services, or other similar means are considered as support. However, normal support work and standard free services are not considered support. Aid that has not been received in cash is assessed and reported in cash.

Grants received at the earliest 2 months before the start of the collection of statements of support and no later than 3 months after the initiative has been submitted to Parliament are considered support.

Grants that cannot be identified are not eligible. However, this does not apply to aid received from normal collection activities.

§ 14

Procedure for notifying aid

The aid must be announced on the computer network in such a way that the information is available to the public no later than the end of the calendar month following the receipt of the aid, but not earlier than the start of the collection of statements of support.

When collecting statements of support, information on the web address where the support is announced must be available to the signatories of the citizens' initiative.

Data on support must be deleted from the data network at the end of the parliamentary consideration of the initiative. If the initiative is not submitted to Parliament, the information must be deleted after the collection of statements of support has ended or the initiative has expired in accordance with Section 11 (2).

§ 15

Appeal

Pursuant to this Act, a decision made by FICORA, the Ministry of Justice, and the Population Register Center may be appealed as provided in the Administrative Procedure Act (586/1996).

§ 16

Penalty provisions

Penalties for giving false personal information and giving false certificates to an authority, as well as for secrecy offenses and breaches of secrecy, as well as for offenses against the personal register, are provided for in the Criminal Code (39/1889).

§ 17

Passage

This Act shall enter into force on 1 March 2012. However, Sections 8 and 9 (2) of the Act shall not enter into force until 1 December 2012.

HE 46/2011 PeVM

6/2011

EV 75/2011

Helsinki, 13 January 2012

President

of the Republic TARJA HALONEN

Minister of Defense

Stefan Wallin

Acts in original

Texts of acts without amendment as published in the collection of acts. All acts from 1987 onwards and a central part of the older acts.

Appendix II

Bias and Social Influence Observation Checklist

Pay active attention to:

Prejudgments

1. Do any of the decision makers stand to personally gain more from one conclusion than another? Yes__ No__
2. Were dissenters encouraged to express their views? Yes__ No__
3. Is there a risk of escalating commitment (sunk costs) to a losing course of action? Yes__ No__

Social Influences

4. Were any of the decision issues recently addressed in the mass media? Yes__ No__
5. Was there any indication that any of the decision issues were previously discussed amongst participants' family members, friends, or colleagues? Yes__ No__
6. Did any decision-maker materially side or identify with any of his or her sources (if any)? Yes__ No__
7. Is there any reason to think that members of the group share any biases, which could lead their errors to be correlated? Yes__ No__
8. Was anyone already committed to a conclusion or is there any reason to suspect prejudice? Yes__ No__

Diversity of Views

9. Were most relevant points of view or expertise represented in this decision-making group? Yes__ No__
10. Did the decision-making group take the outside view for part of its deliberations and seriously attempt (red teamed) so as to apply comparative instead of absolute judgment? Yes__ No__

Availability and Salience

11. Did the participants exaggerate the importance of an event because of its recency, its dramatic quality, or its personal relevance, even if it is not diagnostic? Yes__ No__
12. Is the decision team overly optimistic? Yes__ No__

Substitution

13. Did the group's selection of evidence and the focus of their discussion indicate substitution of easier questions instead of any of the difficult ones assigned? Yes__ No__
14. Did the group neglect or discount an important factor or give more weight to an irrelevant or unimportant one? Yes__ No__

Information Handling

15. Did the judgement rely heavily on anecdotes, stories, or analogies and did available data confirm them? Yes__ No__
16. Did numbers of uncertain accuracy or relevance play a significant role in the final judgment? Yes__ No__
17. Did participants resort to average extrapolations, estimates, or forecasts? Yes__ No__

Decision Process

18. If forecasts were used, did participants question their sources and their validity? Yes__ No__
19. Were confidence intervals (ranges) used for uncertain measurements and were they wide enough? Yes__ No__

20. Is the risk appetite of the decision-makers consistent with that of the organisation? Yes__ No__
21. Is the decision team overly cautious? Yes__ No__
22. Are calculations (including the discounting of the future to present values) consistent with the organisation's long-term strategies and short-term plans? Yes__ No__

Premature Closure/Excessive Coherence

23. Was there undue weight placed in the choice of considerations that were discussed early, or was more time spent on them? Yes__ No__
24. Were alternatives seriously considered, and was evidence that would support them or refute them actively sought? Yes__ No__
25. Were uncomfortable opinions and information supressed or neglected? Yes__ No__
26. Were unintended consequences discussed and considered? Yes__ No__

Scoring and Analysis

Score a 1 if "Yes" for questions 2, 9, 10, 15, 18, 19, 20, 22, 24, and 26 = ____

Score a 1 if "No" for questions 1, 3, 4, 5, 6, 7, 8, 11, 12, 13, 14, 16, 17, 21, 23, 25 = _____

Total observer team "debiased" score = ____

Administer and aggregate the score of a self-assessment version of the checklist. Compare and discuss differences between the decision-group's self-assessment with the bias observation team's results.

Re-assess and reach final decisions.

Adapted from: [82, pp., 388–389].

Appendix III

Narendra Modi: Prime Minister of India - Power, Democracy, War and Peace; Interview with Lex Fridman, 18 March 2025

Lex Fridman: You have a reputation for being a decisive leader. So can you walk me through on this topic of ideas, how you make decisions? What's your process? So, for instance, when facing a high stakes choice with no clear precedence, a lot of uncertainty, having to balance input, how do you make decisions?

Narendra Modi: There are many factors to my decision making. First, I'm perhaps the only politician in India who has stayed overnight in around 85–90% of the districts across the country. This was before my current role. I used to travel extensively. I learned a lot from those experiences. They gave me first-hand knowledge of the ground realities and grassroots level issues, not something asked or heard or learned merely from books. Secondly, from a governance perspective, I carry no baggage of any kind. I don't carry any baggage that weighs me down or forces me to act a certain way. Thirdly, I have a simple yardstick for decisions. My country first. I always question if what I'm doing harms my nation in any way. Additionally, Mahatma Gandhi once said that, if you're ever unsure when making a decision, think of the poorest person's face. Remember them and ask yourself, "Will this help them?" Then your decision will be right. That wisdom always guides me, remembering ordinary citizens and considering how my actions affect them.

Another factor in my approach is that I'm very well-connected in my administration. My officials know this well and probably feel overwhelmed by it, by the fact that my information channels are numerous and are very active. Because of that, I receive lots of insights from various sources. So when someone comes to brief me, that's not my only source of information. I always have additional perspectives available to me.

Another thing, I maintain a learner's mindset. Suppose I'm not familiar with something and an official explains it to me. I approach them like a student and ask, "Can you clarify this? How does it work? Then what happens next and how?" Whenever I have different information, I deliberately play devil's advocate and ask challenging questions. I thoroughly analyze the issue from multiple angles, hoping that careful evaluation will yield something valuable. Then once I converge toward a decision or action that is worth taking, I share the idea informally with like-minded people just to gauge their reactions and see how they respond, gathering insights and feedback before moving forward, until finally, I have a strong conviction that my decision is right.

This entire decision-making process doesn't actually take much time. My speed is very fast. Let me share an example. How did I make decisions during COVID? I had Nobel Prize winners advising me, giving countless economic examples from around the world. They'd say, "This country is doing this, that country did that. You should do it too." Renowned economists constantly bombarded me with suggestions. Political parties pressured me relentlessly, urging me to spend huge amounts of money, but I didn't act immediately. I paused and reflected. What exactly should I do? Then considering the unique conditions of my own country, I made a clear decision. I wouldn't let the poor sleep hungry. I wouldn't allow social tensions to arise over basic daily needs.

These core principles guided my approach. The entire world was in lockdown. Global economies were collapsing. Everyone pressured me to empty the treasury, print more currency and flood money everywhere, but I decided that this was not the right economic route to follow. And so instead, the path I chose, after carefully listening to experts, understanding their opinions without opposing them and combining their advice with my own country's situation and my personal experiences, created a system that worked effectively. As a result, when the whole world suffered from severe inflation immediately after COVID, India did not.

Today, my country is steadily advancing at a rapid pace, emerging as one of the fastest growing major economies in the world. The main reason is that during that crisis, with patience and discipline, I resisted the temptation to apply every global theory blindly. We didn't worry about what newspapers would say, whether they'd praise or criticize. Ignoring all that, I stayed focused on basic fundamentals, and by doing so, we succeeded and kept moving ahead. So ultimately, my economy benefited as well. My approach has always been to stay focused on these fundamentals.

Another strength is my risk-taking capacity. I don't worry about potential losses for myself. If something is right for my country, for the people, I'm always prepared to take the risk. Secondly, I take ownership of my decisions. If something goes wrong, I don't ship blame to others. I stand up, take responsibility, and own the outcome. When you take ownership, your team also becomes deeply committed. They know this person won't let us down, won't abandon us. He'll always stand with us because they see I'm making honest decisions, not for myself, but for the nation. I've openly told the country from the start, I'm human. I can make mistakes, but I won't act with bad intentions. People remember those words clearly. Even if something doesn't go as planned, they trust that Modi's intentions were right. They think he probably meant to do something good even if it didn't work out. So society sees and accepts me just as I am [305].

Bibliography

1. Wikipedia Contributors, "History of Science," Wikipedia, [Online]. Available: https://en.wikipedia.org/wiki/History_of_science. [Accessed 1 September 2019].
2. P. Railton, "Intuitive Guidance: Emotion, Information, and Experience," in *Homo Prospectus*, New York, Oxford University Press, 2016, pp. 33-84.
3. A. Etzioni, "Mixed-Scanning: A "Third" Approach to Decision-Making," *Public Administration Review,* vol. 27, no. 5, pp. 385-392, 1967.
4. A. Huberman, Presenter, *Optimizing Workspace for Productivity, Focus, and Creativity.* [Sound Recording]. Andrew Huberman Lab. 2022.
5. I. McGilchrist, The Master and His Emissary: The Divided Brain and the Making of the Western World, New Haven: Yale University Press, 2009.
6. I. Grossmann, N. M. Weststrate, M. Ardelt, J. P. Brienza, M. Dong, M. Ferrari, M. A. Fournier, C. S. Hu, H. C. Nusbaum and J. Vervaeke, "The Science of Wisdom in a Polarized World: Knowns and Unknowns," *Psychological Inquiry,* vol. 31, no. 2, pp. 103-133, 2020.
7. J. R. Hale, Lords of the Sea: The Epic Story of the Athenian Navy and the Birth of Democracy, London: Penguin Books, 2009.
8. J. C. Scott, Seeing Like a State: How Certain Schemes to Improve the Human Condition Have Failed, New Haven: Yale University Press, 1998.
9. P. Railton, "Morality and Prospection," in *Homo Prospectus*, New York, New York: Oxford University Press, 2016.
10. A. Koestler, The Ghost in the Machine, London: Hutchinson & Co, 1967.
11. Wikipedia Contributors, "Event Horizon Telescope", Wikipedia [Online]. Available: https://en.wikipedia.org/wiki/Event_Horizon_Telescope. [Accessed 11 April 2019].
12. NASA, "NASA Reveals Webb Telescope's First Images of Unseen Universe," NASA, 12 July 2022. [Online]. Available: https://www.nasa.gov/press-release/nasa-reveals-webb-telescope-s-first-images-of-unseen-universe. [Accessed 12 July 2022].
13. A. Murphy Paul, The Extended Mind: The Power of Thinking Outside the Brain, New York: Mariner Books, 2021.
14. C. Kenneally, "How Grammar Changes Perception," *Scientific American,* vol. 329 , no. 4, pp. 48-59, 2023.
15. World Science Festival, "When Thoughts Aren't Private: Will AI Soon Read Our Minds?," YouTube, 14 March 2025. [Online]. Available: https://www.youtube.com/watch?v=HX0C3qxZggY&t=1725s. [Accessed 6 April 2025].
16. V. Ramachandran, The Tell-Tale Brain: Unlocking the Mystery of Human Nature, London: William Heinemann, 2011.

E. Bellos, *Sustainability Thinking*, https://doi.org/10.1007/978-3-032-11244-6

17. M. Gell-Mann, The Quark and the Jaguar: Adventures in the Simple and the Complex, London: Abacus, 1994, pp. 119-120.
18. B. Hays, "Climate Change to Make Earth More Blue," UPI, 4 February 2019. [Online]. Available: https://www.upi.com/Science_News/2019/02/04/Climate-change-to-make-Earth-more-blue/1071549291285/#:~:text=Feb.%204%20%28UPI%29%20%2D%2D%20Some%20might%20argue%20the,Earth%27s%20oceans%20warm%2C%20they%20will%20become%20more%20blue. [Accessed 17 April 2023].
19. A. Grant, Think Again: The Power of Knowing What you Don't Know, London: WH Allen, 2021.
20. T. Widmer, "What Did Plato Think the Earth Looked Like? For millenniums, humans have tried to imagine the world in space. Fifty years ago, we finally saw it.," New York Times, 2018 December 2018. [Online]. Available: https://www.nytimes.com/2018/12/24/opinion/plato-earth-christmas-eve-apollo-8.html. [Accessed 9 May 2019].
21. Wikipedia Contributors, "Earthrise," Wikipedia, [Online]. Available: https://en.wikipedia.org/wiki/Earthrise#cite_note-NYT-20181224b-14. [Accessed 9 May 2019].
22. P. Tetlock and D. Gardner, Superforecasting: The Art and Science of Prediction, London: Random House Books, 2015.
23. E. O. Wilson, Consilience: The Unity of Knowledge, London: Abacus, 1998.
24. G. Mulgan, Big Mind: How Collective Intelligence Can Change Our World, Princeton, New Jersey: Princeton University Press, 2018.
25. J. LeDoux, The Deep History of Ourselves: The Four-Billion-Year Story of How We Got Conscious Brains, New York: Penguin Books, 2020, pp. 232, 247.
26. Y. N. Harari, Sapiens: A Brief History of Humankind, London: Vintage, 2011.
27. M. Tomasello, A Natural History of Human Thinking, Cambridge, Massachusetts: Harvard University Press, 2014.
28. N. A. Christakis, Blueprint: The Evolutionary Origins of a Good Society, New York, New York: Little, Brown Spark, 2019.
29. B. M. Seitz, A. Aktipis, D. M. Buss, J. Alcock, P. Bloom, M. Gelfand, S. Harris, D. Lieberman, B. N. Horowitz, S. Pinker, D. S. Wilson and M. G. Haselton, "The Pandemic Exposes Human Nature: 10 Evolutionary Insights," *Proceedings of the National Academy of Sciences of the United States of America,* vol. 117, no. 45, pp. 27767-27776, 2020.
30. Wikipedia Contributors, "Noosphere," Wikipedia, [Online]. Available: https://en.wikipedia.org/wiki/Noosphere. [Accessed 21 February 2021].
31. G. West, Scale: The Universal Laws of Life, Growth, and Death in Organisms, Cities, and Comanies, New York: Penguin Books, 2017, pp. 412-426.
32. S. E. Page, The Model Thinker: What You Need to Know to Make Data Work for You, New York: Basic Books, 2018, pp. 32, 147, 336-337.
33. N. N. Taleb, Antifragile: Things that Gain from Disorder, New York: Random House, 2012.
34. G. McGhee, "The Geometry of Evolution: Adaptive Landscapes and Theoretical Morphospaces," Cambridge University Press, Cambridge, 2007.
35. S. G. Tzafestas, Energy, Information, Feedback, Adaptation, and Self-Organization, Cham: Springer International Publishing AG, 2018.
36. J. H. Miller and S. E. Page, Complex Adaptive Systems: An Introduction to Computational Models of Social Life, Princeton: Princeton University Press, 2007, p. 216.
37. O. R. Young, The Institutional Dimensions of Environmental Change: Fit, Interplay, and Scale, Cambridge: MIT Press, 2002.
38. R. Garud and P. Karnoe, Eds., Path Dependence and Creation, Mahwah, NJ: Lawrence Erlbaum Associates, Inc., 2001.
39. S. J. Liebowitz and S. E. Margolis, "Path Dependence," 1999. [Online]. Available: https://reference.findlaw.com//lawandeconomics/0770-path-dependence.pdf. [Accessed 30 November 2020].
40. S. E. Page, Diversity and Complexity, Princeton: Princeton University Press, 2011, p. 26.
41. J. H. Miller and S. E. Page, Complex Adaptive Systems: An Introduction to Computational Models of Social Life, Princeton: Princetion University Press, 2007.

42. M. L. Wong, C. E. Cleland, D. Arend Jr, S. Bartlett, J. H. Cleaves, H. Demarest, A. Prabhu, J. I. Lunine and R. M. Hazen, "On the Roles of Function and Selection in Evolving Systems," *Proceedings of the National Academy of Sciences,* vol. 120, no. 43 , p. e2310223120, 2023.
43. D. Bohm, On Dialogue, London: Routledge, 1996.
44. D. Cabrera, "52nd Annual Meeting of the ISSS," in *Distinctions, Systems, Relationships, Perspectives: The Simple Rules of Complex Conceptual Systems*, Madison, 2008.
45. E. Vasileiadou and K. Safarzynska, "Transitions: Taking Complexity Seriously," *Futures,* vol. 42, pp. 1176-1186, 2010.
46. T. Hämäläinen, "How to deal with growing complexity and uncertainty How can decision-makers deal with increasingly complex and uncertain systems?," SITRA, 14 August 2015. [Online]. Available: https://www.sitra.fi/en/blogs/weekly-notes-week-33-how-deal-growin-complexity-and-uncertainty-0/. [Accessed 16 April 2020].
47. G. C. Gallopin, "Planning for Resilience: Scenarios, Surprises, and Branch Points," in *Panarchy: Understanding Transformations in Human and Natural Systems*, London, Island Press, 2002.
48. R. Preiser, R. Biggs, A. De Vos and C. Folke, "Social-Ecological Systems as Complex Adaptive Systems: Organizing Principles for Advancing Research Methods and Approaches," *Ecology and Society,* vol. 23, no. 4(46), 2018.
49. H. Landemore, Open Democracy: Reinventing Popular Rule for the Twenty-First Century, Princeton: Princeton University Press, 2020.
50. C. Schill, J. M. Anderies, T. Lindahl, C. Folke, S. Polasky, J. C. Cárdenas, A.-S. Crépin, M. A. Janssen, J. Norberg and M. Schlüter, "A More Dynamic Understanding of Human Behaviour for the Anthropocene," *Nature Sustainability,* vol. 2, pp. 1075-1082, 2019.
51. C.S. Holling, L.H. Gunderson and G.D. Peterson, "Sustainability and Panarchies," in *Panarchy: Understanding Transformations in Human and Natural Systems*, Washington, DC: Island Press, 2002.
52. "Adaptive Cycle," Resiliance Alliance, [Online]. Available: https://www.resalliance.org/adaptive-cycle. [Accessed 2 May 2021].
53. L. J. Pearson and C. J. Pearson, "Societal Collapse or Transformation, and Resilience," *Proceedings of the National Academy of Sciences,* vol. 109, no. 30, pp. 2030-2031, 2012.
54. E. Ostrom, "A General Framework for Analyzing Sustainability of Social-Ecological Systems," *Science,* vol. 325, pp. 419-422, 2009.
55. S.M. Sundstrom and C.R. Allen, "The Adaptive Cycle: More Than a Metaphor," *E cological Complexity,* vol. 39, no. 100767, 2019.
56. T. Dietz, E. Ostrom and C. Stern, "The Struggle to Govern the Commons," *Science,* vol. 302, pp. 1907-1912, 2003.
57. R. N. Goldstein, Plato at the Googleplex: Why Philosophy Won't Go Away, London: Atlantic Books, 2014.
58. L. Boroditsky, "How Languages Construct Time," in *Space, Time, and Number in the Brain*, Academic Press, 2011, pp. 333-341.
59. V. Smil, Growth: From Microorganisms to Megacities, Cambridge, Massachusetts: The MIT Press, 2019, p. xiv.
60. J. S. Downs, "Prescriptive scientific narratives for communicating usable science," *Proceedings of the National Academy of Sciences,* vol. 111, pp. 13627-13633, 2014.
61. J. M. Wyndham and M. W. Vitullo, "Define the human right to science," *Science,* vol. 362, no. 6418, p. 975, 2018.
62. C. E. Kearns, L. A. Schmidt and S. A. Glantz, "Sugar industry and coronary heart disease research: A historical analysis of internal industry documents," *JAMA Intern Med.,* vol. 176, no. 11, pp. 1680-1685, 2016.
63. H. Pearson, "The Future of Science Advice," *Nature,* vol. 636, no. 5 December 2024, pp. 26-30, 2024.
64. L. McIntyre, Post-truth, Cambridge: The MIT Press, 2018, p. 53.
65. P. Schmid and C. Betsch, "Effective Strategies for Rebutting Science Denialism in Public Discussions," *Nature Human Behaviour,* vol. 3, no. 9, pp. 931-939, 2019.

66. World Health Organization, Regional Office for Europe, "Best Practice Guidance: How to Respond to Vocal Vaccine Deniers in Public," WHO, Copenhagen, 2016.
67. R. K. Merton, "The Matthew effect in science," *Science,* vol. 159, no. 3810, pp. 55-63, 1968.
68. Q. Ke, E. Ferrara and F. F. A. Radicchi, "Defining and identifying sleeping beauties in science," *Proceedings of the National Academy of Sciences,* vol. 112, no. 24, pp. 7426-7431, 2015.
69. A. Corner, C. Shaw and J. Clarke, "Principles for Effective Communication and Public Engagement on Climate Change: A Handbook for IPCC authors," Oxford, Climate Outreach, 2018.
70. United Nations Department of Economic and Social Affairs, "The Sustainable Development Goals Report: 2022," UN, 2022.
71. E. Tenner, The Efficiency Paradox: What Big Data Can't Do, New York: Alfred A. Knopf, 2018, pp. 52-53.
72. J. Lovelock, Gaia: A New Look at Life on Earth, Oxford: Oxford University Press, 2000.
73. P. A. Corning, S. A. Kauffman, D. Noble, J. A. Shapiro, R. I. Vane-Wright and A. Pross, "Introduction," in *Evolution "On Purpose": Teleonomy in Living Systems*, The MIT Press, 2023, pp. 1-9.
74. P. A. Corning, "Teleonomy in Evolution: "The Ghost in the Machine"," in *Evolution "On Purpose": Teleonomy in Living Systems*, The MIT Press, 2023, pp. 11-31.
75. B. Latour, Down to Earth: Politics in the New Climate Regime, Cambridge, UK: Polity Press, 2018.
76. Y. N. Harari, Nexus: A Brief History of Information Networks from the Stone Age to AI, London: Fern Press, 2024.
77. H. A. Kissinger, "How the Enlightenment Ends," The Atlantic, 1 June 2018. [Online]. Available: https://www.henryakissinger.com/articles/how-the-enlightenment-ends/. [Accessed 12 December 2021].
78. M. Gleiser, The Island of Knowledge: The Limits of Science and the Search for Meaning, New York: Basic Books, 2014.
79. R. K. Merton, "The Matthew Effect in Science," *Science,* vol. 159, no. 3810, pp. 56-63, 1968.
80. A. Neuberger, N. Oraiopoulos and D. L. Drakeman, "Limits of Drug Repurposing," University of Cambridge Judge Business School, 13 August 2018. [Online]. Available: https://www.jbs.cam.ac.uk/2018/limits-of-drug-repurposing/. [Accessed 29 December 2023].
81. T. Lombrozo, "Learning by Thinking," Edge Foundation, [Online]. Available: https://www.edge.org/print/conversation/tania_lombrozo-learning-by-thinking. [Accessed 14 1 2022].
82. D. Kahneman, O. Sibony and C. R. Sunstein, Noise: A Flaw in Human Judgment, London: William Collins, 2021, p. 40.
83. Wikipedia Contributos, "List of Cognitive Biases," Wikipedia [Online]. Available: https://en.wikipedia.org/wiki/List_of_cognitive_biases. [Accessed 17 January 2019].
84. S. Pinker, Rationality: What it is, Why it Seems Scarce, Why it Matters, London: Allen Lane, 2021.
85. D. Kahneman, Thinking, Fast and Slow, London: Penguin Books, 2011, pp. 212, 252.
86. C. Stripada, "Deliberative Guidance: Intuitive Guidance in Counterfactual Mode," in *Homo Prospectus*, New York, Oxford University Press, 2016, pp. 87-101.
87. M. Zenko, Red Team: How to Succeed by Thinking Like the Enemy, New York: Basic Books, 2015, pp. 210-212.
88. M. Seligman, M. Forgeard and S. B. Kaufman, "Creativity and Aging: What We Can Make With What We Have Left," in *Homo Prospectus*, New York, Oxford University Press, 2016, pp. 305-350.
89. J. A. Dearing, "Human-Environment Interactions: Learning from the Past," in *Sustainability or Collapse: An Integrated History and Future of People on Earth*, Cambridge, The MIT Press, 2007, pp. 19-37.
90. B. J. M. de Vries, "Scenarios: Guidance for an Uncertain and Complex World?," in *Sustainability and Collapse: An Integrated History and Future of People on Earth*, Cambridge, The MIT Press, 2007, pp. 379-397.
91. D. Sarewitz, "Saving Science," *The New Atlantis,* vol. Spring/Summer, 2016.

92. M. Rees, On the Future Prospects for Humanity, Princeton: Princeton University Press, 2018, pp. 209-210.
93. D. Sarewitz, "Saving Science," *The New Atlantis,* no. Spring/Summer, pp. 5-40, 2016.
94. J. Ryan, "Plagiarism in Peer-Review Reports Could be the 'Tip of the Iceberg'," Nature Index, 1 May 2024. [Online]. Available: https://www.nature.com/articles/d41586-024-01312-0. [Accessed 6 May 2024].
95. A. Swan, "Policy Guidelines for the Development and Promotion of Open Access," United Nations Educational, Scientific and Cultural Organization, Paris, 2012.
96. Organisation for Economic Co-operation and Development, "Systems Approaches to Public Sector Challenges: Working with Change," OECD Publishing, Paris, 2017.
97. J. M. Polimeni, K. Mayumi, M. Giampietro and B. Alcott, "The Jevons Paradox: The Evolution of Complex Adaptive Systems and the Challenge for Scientific Analysis," in *The Myth of Resource Efficiency: The Jevons Paradox*, New York, Earthscan, 2008, pp. 79-140.
98. M. Stafford-Smith, O. Gaffney, L. Brito, E. Ostrom and S. Seitzinger, "Interconnected Risks and Solutions for a Planet Under Pressure– Overview and Introduction," *Current Opinion in Environmental Sustainability,* no. 4, pp. 3-6, 2012.
99. C. Folke, S. Polasky, J. Rockström, V. Galaz, F. Westley, M. Lamont, M. Scheffer, H. Österblom, S. R. Carpenter, F. Stuart Chapin III, K. C. Seto and E. Weber, "Our Future in the Anthropocene Biosphere," *Ambio,* vol. 50, pp. 834-869, 2021.
100. O. Renn, Risk Governance: Coping with Uncertainty in a Complex World, London: Earthscan, 2008.
101. N. Bostrom, "Existential Risks: Analyzing Human Extinction Scenarios and Related Hazards," *Journal of Evolution and Technology,* vol. 9, 2002.
102. N. Bostrom, "Existential Risk Prevention as Global Priority," *Global Policy,* vol. 4, no. 1, pp. 15-31, 2013.
103. E. Bellos, "Sustainable Energy Development: How Can the Tension Between Energy Security and Energy Transition be Measured and Managed in South Africa?," *Journal of Cleaner Production,* vol. 205, pp. 738-753, 2018.
104. International Court of Justice, "Obligations of States In Respect of Climate Change," General List No. 187, 2025.
105. I. Linkov, T. Bridges, F. Creutzig, J. Decker, C. Fox-Lent, W. Kröger and J. H. Lambert, "Changing the Resilience Paradigm," *Nature Climate Change,* vol. 4, no. 6, pp. 407-409, 2014.
106. I. Linkov, D. A. Eisenberg, M. E. Bates, D. Chang, M. Convertino, J. H. Allen, F. S. E. and T. P. Seager, "Measurable Resilience for Actionable Policy," *Environmental Science and Technology* , vol. 47, p. 10108–10110, 2013.
107. S. Wood, A. Luers, J. Garard, A. Gambhir, K. Chaudhari, M. Ivanova and C. Cronin, "Collective Foresight and Intelligence for Sustainability," *Global Sustainability,* vol. 4, 2021.
108. K. Kupferschmidt, "Global Pandemic Treaty Finalized, Without U.S., in "a Victory for Multilateralism"," *Scienceinsider,* pp. 1-7, 16 April 2025.
109. Wikipedia Contributors, "Doomsday Clock," Wikipedia, [Online]. Available: https://en.wikipedia.org/wiki/Doomsday_Clock. [Accessed 29 November 2020].
110. Science and Security Board Bulletin of the Atomic Scientists, "A Time of Unprecedented Danger: It is 90 Seconds to Midnight," 24 January 2023. [Online]. Available: https://thebulletin.org/doomsday-clock/current-time/2023. [Accessed 27 January 2023].
111. J. Mecklin, "Closer Than Ever: It Is Now 89 Seconds To Midnight," *Bulletin of the Atomic Scientists,* pp. 1-9, 28 January 2025.
112. S. McChrystal and A. Butrico, Risk: A User's Guide, London: Penguin Random House UK, 2021.
113. J. Liu, H. Mooney, V. Hull, S. Davis, J. Gaskell, T. Hertel, J. Lubchenco, K. Seto, P. Gleick, C. Kremen and S. Li, "Systems Integration for Global Sustainability," *Science,* vol. 347, no. 1258832, 2015.
114. C. Holling, S. R. Carpenter, W. A. Brock and L. H. Gunderson, "Discoveries for Sustainable Futures," in *Panarchy: Understanding Transformations in Human and Natural Systems,* Washington, Island Press, 2002.

115. F. Berkes and H. Ross, "Panarchy and Community Resilience: Sustainability Science and Policy Implications," *Environmental Science & Policy,* vol. 61, pp. 185-193 , 2016.
116. M. Stafford-Smith, O. Gaffney, L. Brito, E. Ostrom and S. Seitzinger, "Interconnected Risks and Solutions for a Planet Under Pressure – Overview and Introduction," *Current Opinion in Environmental Sustainability,* vol. 4, pp. 3-6, 2012.
117. D. Meadows, Leverage Points: Places to Intervene in a System, Hartland, Vermont: Sustainability Institute , 1999.
118. United Nations Department of Economic and Social Affairs, "The 17 Goals," United Nations, [Online]. Available: https://sdgs.un.org/goals. [Accessed 29 November 2020].
119. United Nations Statistical Commission, "SDG Indicators," [Online]. Available: https://unstats.un.org/sdgs/indicators/indicators-list/. [Accessed 9 May 2021].
120. Y. Zeng, S. Maxwell, R. K. Runting, O. Venter, J. E. M. Watson and L. R. Carrasco, "Environmental Destruction Not Avoided With the Sustainable Development Goals," *Nature Sustainability,* vol. 3, no. 10, p. 795–798. , 2020.
121. United Nations Department of Economic and Social Affairs, "The Sustainable Development Goals Report 2024," United Nations, New York, 2024.
122. C. Epstein, "Common but Differentiated Responsibilities: International Environmental Law," Encyclopedia Britannica, [Online]. Available: https://www.britannica.com/topic/common-but-differentiated-responsibilities. [Accessed 14 December 2019].
123. United Nations Environment Programme, "Global Environmental Outlook: GEO-6, Healthy Planet, Healthy People," Cambridge University Press, New York, 2019.
124. P. R. Bixler, S. Atshan, J. L. Banner, D. Tremaine and R. E. Mace, "Assessing Integrated Sustainability Research: Use of Social Network Analysis to Evaluate Scientific Integration and Transdisciplinarity in Research Networks," *Current Opinion in Environmental Sustainability,* vol. 39, pp. 103-113, 2019.
125. J. Gottschall, The Storytelling Animal: How Stories Make us Human, New York: First Mariner Books, 2013, p. 8.
126. E. Ostrom, Governing the Commons: The Evolution of Institutions for Collective Action, Cambridge: Cambridge University Press, 1990.
127. R. Beddoe, R. Costanza, J. Farley, E. Garza, J. Kent, I. Kubiszewski, L. Martinez, T. McCowen, K. Murphy, N. Meyers, Z. Ogden, K. Stapleton and J. Woodward, "Overcoming Systemic Roadblocks to Sustainability: The Evolutionary Redesign of Worldviews, Institutions, and Technologies," *Proceedings of the National Academy of Sciences,* vol. 106, no. 8, pp. 2483-2489, 2009.
128. Working Group on the 'Anthropocene', "Results of Binding Vote by AWG," Subcommission on Quaternary Stratigraphy , 21 May 2019. [Online]. Available: http://quaternary.stratigraphy.org/working-groups/anthropocene. [Accessed 12 March 2021].
129. S. L. Lewis and M. A. Maslin, "Defining the Anthropocene," *Nature,* vol. 519, pp. 171-180, 2015.
130. W. Steffen, J. Rockström, K. Richardson, T. M. Lenton, C. Folke, D. Liverman, C. P. Summerhayes, et. al., "Trajectories of the Earth System in the Anthropocene," *Proceedings of the National Academy of Sciences,* vol. 115, no. 33, pp. 8252-8259, 2018.
131. W. Steffen, J. Grinevald, P. Crutzen and J. McNneill, "The Anthropocene: Conceptual and Historical Perspectives," *Philosophical Transactions of the Royal Society,* vol. 369, pp. 842-867, 2011.
132. E. Elhacham, L. Ben-Uri, J. Grozovski, Y. M. Bar-On and R. Milo, "Global Human-made Mass Exceeds All Living Biomass," *Nature,* vol. 58, no. 7838, pp. 442-444, 2020.
133. Aspen Global Change Institute, "The Anthroposphere," [Online]. Available: https://www.agci.org/earth-systems/anthroposphere. [Accessed 15 April 2024].
134. J. Rockström, W. Steffen, K. Noone, Å. Persson, F.S. Chapin III, E. Lambin, T.M. Lenton, M. Scheffer, C. Folke, H.J. Schellnhuber and B. Nykvist, "Planetary Boundaries: Exploring the Safe Operating Space for Humanity," *Ecology and Society,* vol. 14(2), no. 32, 2009.
135. W. Steffen, K. Richardson, J. Rockström, S. E. Cornell, I. Fetzer, E. M. Bennett and R. Biggs, "Planetary Boundaries: Guiding Human Development on a Changing Planet," *Science,* vol. 347, no. 6223, p. 1259855, 2015.

136. J. Rockström, J. Gupta, D. Qin, S. J. Lade, J. F. Abrams, L. S. Andersen, D. I. Armstrong McKay, X. Bai, G. Bala, S. E. Bunn and D. Ciobanu, "Safe and Just Earth System Boundaries," *Nature*, 619(7968), pp.102-111, 2023.
137. K. Richardson, W. Steffen, W. Lucht, J. Bendtsen, S.E. Cornell, J.F. Donges, M. Drüke, I. Fetzer, G. Bala, W. Von Bloh and G. Feulner, "Earth Beyond Six of Nine Planetary Boundaries," *Science Advances,* vol. 9, no. 37, p. eadh2458, 2023.
138. The Futures Group International, "Decision Modeling," Futures Research Methods v3.0, The Millennium Project, 2009.
139. J.-J. Wang, Y.-Y. Jing, C.-F. Zhang and J.-H. Zhao, "Review on Multi-Criteria Decision Analysis Aid in Sustainable Energy Decision-Making," *Renewable and Sustainable Energy Reviews,* vol. 13, no. 9, pp. 2263-2278, 2009.
140. J. Diamond, Collapse: How Societies Choose to Fail or Survive, London: Penguin Books, 2005.
141. Leach, Melissa, Belinda Reyers, Xuemei Bai, Eduardo S. Brondizio, Christina Cook, Sandra Díaz, Giovana Espindola, Michelle Scobie, Mark Stafford-Smith, and Suneetha M. Subramanian. "*Equity and sustainability in the Anthropocene: A social–ecological systems perspective on their intertwined futures.*" *Global Sustainability* 1 (2018): e13.
142. F. Biermann, "The Case for a World Environment Organization," *Environment: Science and Policy for Sustainable Development,* vol. 42, no. 9, pp. 22-31, 2000.
143. J. Ober, "Democracy's Wisdom: An Aristotelian Middle Way for Collective Judgment," *American Political Science Review,* vol. 107, no. 1, pp. 104-122, 2013.
144. M. New, D. Viner, C. Adler, S.-M. Cheong, C. Conde, A. Constable, E. Coughlan de Perez, A. Lammel, R. Mechler, B. Orlove and W. Solecki, "Decision-Making Options for Managing Risk," in *Climate Change 2022: Impacts, Adaptation and Vulnerability*, Cambridge, Cambridge University Press, 2022, pp. 2539–2654.
145. P. M. Senge, The Fifth Discipline: The Art and Practice of the Learning Organization, London: Random House Business Books, 2006.
146. C. List and R. E. Goodin, "Epistemic Democracy: Generalizing the Condorcet Jury Theorem," *Journal of Political Philosophy,* vol. 9, no. 3, 2001.
147. S. Sloman and P. Fernbach, The Knowledge Illusion: The Myth of Individual Thought and the Power of Collective Wisdom, London: Pan Books, 2018.
148. N. A. Christakis, Blueprint: The Evolutionary Origins of a Good Society, New York: Little, Brown Spark, 2019.
149. J. Kuyper, "Global Democracy," The Stanford Encyclopedia of Philosophy, 6 October 2016. [Online]. Available: https://plato.stanford.edu/archives/win2016/entries/global-democracy/. [Accessed August 2022].
150. D. M. Estlund, Democratic Authority: A Philosophical Framework, Princeton: Princeton University Press, 2008.
151. J. D. Erickson, The Progress Illusion: Reclaiming Our Future from the Fairytale of Economics, Washington: Island Press, 2022.
152. L. Hong and S. E. Page, "Some Microfoundations of Collective Wisdom," in *Collective Wisdom: Principles and Mechanisms*, New York, Cambridge University Press, 2012, pp. 56-71.
153. R. Thaler and C. Sunstein, Nudge: Improving Decisions About Health, Wealth, and Happiness, London: Penguin Books, 2009.
154. K. Yoder, "A Common Talking Point About Climate Change Gets It All Wrong, New Study Says," Grist, 21 April 2023. [Online]. Available: https://grist.org/language/climate-talking-point-wrong-psychological-distance-study/. [Accessed 28 September 2023].
155. L. Smith and A. Archer, "Epistemic Injustice and the Attention Economy," *Ethical Theory and Moral Practice,* vol. 23, no. 5, pp. 777-795, 2020.
156. T. R. Miller, A. Wiek, D. Sarewitz, J. Robinson, L. Olsson, D. Kriebel and D. Loorbach, "The Future of Sustainability Science: A Solutions-Oriented Research Agenda," *Sustainability Science,* vol. 9, pp. 239-246, 2014.
157. N. Bostrom, "Information Hazards: A Typology of Potential Harms from Knowledge," *Review of Contemporary Philosophy,* vol. 10, pp. 44-79, 2011.

158. Nature Editorial, "The World's Goals for Saving Humanity Are Still the Best Option," *Nature,* vol. 621, pp. 227-229, 2023
159. Nature Editorial, ""Hidden" Science Can Help Tackle the Biggest Problems," *Nature,* vol. 633, p. 493, 2024.
160. United Nations, "Summit of the Future," in *Pact for the Future, Global Digital Compact and Declalation on Future Generations*, New York, 2024.
161. M. Pelacho, H. Rodríguez, F. Broncano, R. Kubus, F. Sanz García, B. Gavete and A. Lafuente, "Science as a Commons: Improving the Governance of Knowledge Through Citizen Science," in *The Science of Citizen Science*, Cham, Springer, 2021, pp. 57-78.
162. E. Tenner, The Efficiency Paradox: What Big Data Can't Do, New York: Albert A Knopf, 2018, p. 121.
163. D. Sarewitz, "Saving Science," *The New Atlantis,* vol. Spring/Summer , pp. 5-40, 2016.
164. P. Railton, "Introduction," in *Homo Prospectus*, New York, Oxford University Press, 2016, pp. 3-31.
165. Wikipedia Contributors, "Self Categorization Theory," Wikipedia, [Online]. Available: https://en.wikipedia.org/w/index.php?title=Self-categorization_theory&oldid=1173760429. [Accessed 4 October 2023].
166. J. Mendelsohn, "Aristotle: Epistemology," Internet Encyclopedia of Philosophy, [Online]. Available: https://iep.utm.edu/aristotle-epistemology/. [Accessed 9 September 2025].
167. S. G. Tzafestas, Energy, Information, Feedback, Adaptation, and Self-Organization: The Fundamental Elements of Life and Society, Cham: Springer International Publishing AG, 2018, p. 481.
168. R. Baumeister, "Collective Prospection: The Social Construction of the Future," in *Homo Prospectus*, New York, Oxford University Press, 2016.
169. H. Václav, "The Power of the Powerless," International Center on Nonviolent Conflict, October 1978. [Online]. Available: htpss://www.nonviolent-conflict.org/resource/the-power-of-the-powerless-book/. [Accessed 2020 July 28].
170. J. Ober, "Democracy's Wisdom: An Aristotelian Middle Way for Collective Judgement," *American Political Science Review,* vol. 107, no. 1, pp. 104-122, 2013.
171. N. N. Taleb, Skin in the Game: Hidden Asymmetries in Daily Life, London: Allen Lane, 2018.
172. G. Quaggiotto, B. Leurs, Christiansen and Jesper, "Towards an Experimental Culture in Government: Reflections On and From Practice," Nesta, [Online]. Available: https://www.nesta.org.uk/blog/towards-an-experimental-culture-in-government-reflections-on-and-from-practice/. [Accessed 13 12 2021].
173. M. Annala, T. Kaskinen, S. Lee, J. Leppänen, K. Mattila, A. Neuvonen, J. Nuutinen, E. Saarikoski and A. Tarvainen, "Design For Government: Human-Centric Governance Through Experiments," Experimental Finland, Helsinki, 2015.
174. E. Kandel, In Search of Memory: The Emergence of a New Science of Mind, New York: W. W. Norton & Company, 2006.
175. P. Railton, "Intuitive Guidance: Emotion, Information, and Experience," in *Homo Prospectus*, New York, Oxford University Press, 2016, p. 63.
176. J. De Freitas, K. Thomas, P. DeScioli and S. Pinker, "Common Knowledge, Coordination, and Strategic Mentalizing in Human Social Life," *Proceedings of the National Academy of Sciences,* vol. 116, no. 28, pp. 13751-13758 , 2019.
177. D. P. Redlawsk, A. J. Civettini and K. M. Emmerson, "The Affective Tipping Point: Do Motivated Reasoners Ever "Get it"?," *Political Psychology,* vol. 31, no. 4, pp. 563-593, 2010.
178. J. Kalla and D. Broockman, "The Minimal Persuasive Effects of Campaign Contact In General Elections: Evidence From 49 Field Experiments," *American Political Science Review* , vol. 112, no. 1, pp. 148-166, 2018.
179. D. McRaney, How Minds Change: The New Science of Belief, Opinion and Persuasion, London: Oneworld, 2022.
180. A. Yang, "Data as a Property Right," Yang 2020, [Online]. Available: https://www.yang2020.com/policies/data-property-right/. [Accessed 2 October 2019].

181. G. Mackie, "Rational Ignorance and Beyond," in *Collective Wisdom: Principles and Mechanisms*, New York, Cambridge University Press, 2012.
182. C. Crimston, H. Selvanathan and J. Jetten, "Moral Polarization Predicts Support for Authoritarian and Progressive Strong Leaders via the Perceived Breakdown of Society," *Political Psychology,* 2021.
183. G. Pennycook, Z. Epstein, M. Mosleh, A. A. Arechar and D. R. D. G. Eckles, "Shifting Attention to Accuracy Can Reduce Misinformation Online," *Nature,* vol. 592, pp. 590-595, 2021.
184. H. R. Markus and S. Kitayama, "Cultures and Selves: A Cycle of Mutual Constitution," *Perspectives on Psychological Science,* vol. 5, no. 4, pp. 420-430, 2010.
185. Wikipedia Contributors, "Hebbian Theory," Wikipedia [Online]. Available: https://en.wikipedia.org/wiki/Hebbian_theory. [Accessed 10 October 2023].
186. J. Ober, "Epistemic Democracy in Classical Athens: Sophistication, Diversity, and Innovation," in *Collective Wisdom: Principles and Mechanisms*, New York, Cambridge University Press, 2014, pp. 118-147.
187. C. Stripada, "Free Will and the Construction of Options," in *Homo Prospectus*, New York, Oxford University Press, 2016, pp. 191-206.
188. M. Seligman, "Afterword," in *Homo Prospectus*, New York, Oxford University Press, 2016, pp. 351-355.
189. N. Oxman, "Age of Entanglement," Journal of Design and Science, 2016. [Online]. Available: https://www.pubpub.org/pub/ageofentanglement?context=jods. [Accessed 22 July 2023].
190. L.-M. Glass and J. Newig, "Governance for Achieving the Sustainable Development Goals: How Important are Participation, Policy Coherence, Reflexivity, Adaptation and Democratic Institutions?," *Earth System Governance,* vol. 2, no. 100031, pp. 1-14, 2019.
191. United Nations Department of Economic and Social Affairs, "SDG Indicators Metadata Repository," Statistics Division, [Online]. Available: https://unstats.un.org/sdgs/metadata?Text=&Goal=4&Target=4.7. [Accessed 23 July 2023].
192. United Nations Educational, Scientific and Cultural Organization, "Education for Sustainable Development: A Roadmap," UNESCO, Paris, 2020.
193. Our World in Data Team , "Ensure Inclusive and Quality Education for All and Promote Lifelong Learning," OurWorldInData.org, 2023. [Online]. Available: https://ourworldindata.org/sdgs/quality-education. [Accessed 19 July 2023].
194. P. Francis, *"Laudato Si.": On Care for Our Common Home,* Vatican City: Vatican Press, 2015.
195. C. S. Holling and L. H. Gunderson, "Resilience and Adaptive Cycles," in *Panarchy: Understanding Transformations in Human and Natural Systems*, Washington, Island Press, 2002, pp. 25-62.
196. K. Raworth, Doughnut Economics: Seven Ways to Think Like a 21st-Century Economist, White River Junction: Chelsea Green Publishing, 2017.
197. Y. Y. Haimes, "Systems-based Risk Analysis," in *Global Catastrophic Risks*, Oxford, Oxford University Press, 2008, pp. 146-163.
198. K. Dorst, Frame Innovation: Create New Thinking by Design, Cambridge: The MIT Press, 2015.
199. J. Gottschall, The Storytelling Animal: How Stories Make Us Human, New York: Mariner Books, 2013.
200. J. Haidt, The Happiness Hypothesis: Putting Ancient Wisdom and Philosophy to the Test of Modern Science, London: Arrow Books, 2006.
201. H.-O. Pörtner, D. Roberts, E. Poloczanska, K. Mintenbeck, M. Tignor, A. Alegría, M. Craig, S. Langsdorf, S. Löschke, V. Möller, A. Okem and B. Rama, "Climate Change 2022: Impacts, Adaptation and Vulnerability," Cambridge University Press, Cambridge, 2022.
202. R. Lengler and M. J. Eppler, "Towards a Periodic Table of Visualization Methods for Management," *IASTED Proceedings of the Conference on Graphics and Visualization,* vol. 7, no. 8, p. 31, January 2007.
203. C. Sanderson, The Bystander Effect: The Psychology of Courage and How to Be Brave, London: William Collins, 2020.

204. E. F. Barkley and C. H. Major, Student Engagement Techniques: A Handbook for College Faculty, Hoboken, NJ: John Wiley & Sons, 2020.
205. Santa Fe Institute, "Science Board & Trustees Symposium on the Emergent Political Economy of Climate Change," [Online]. Available: https://santafe.edu/events/science-board-trusteessymposium-emergent-political-economy-climate-change. [Accessed 2 June 2022].
206. Nature Editorial, "Concrete Needs to Lose its Colossal Carbon Footprint," *Nature* [Online]. Available: https://www.nature.com/articles/d41586-021-02612-5. [Accessed 6 June 2022]
207. National Academies of Sciences, Engineering, and Medicine, "Communicating Science Effectively: A Research Agenda," National Academies of Sciences, Engineering, and Medicine, 2017.
208. R. Todd and J. Lasky-Fink, "Harvard Researchers Explain How Our Brains Skim Information and How to Actually Get People's Attention," Fast Company, 12 September 2023. [Online]. Available: https://www.fastcompany.com/90951313/harvard-researchers-explain-how-our-brains-skim-information-and-how-to-actually-get-peoples-attention. [Accessed 26 September 2023].
209. S. Tamm, "Feynman Technique: A Complete Beginner's Guide," e-student.org , 24 May 2021. [Online]. Available: https://e-student.org/feynman-technique/. [Accessed 17 April 2022].
210. United Nations Environment Programme, "Global Environment Outlook (GEO-6): Healthy Planet, Healthy People," UNEP, Nairobi, 2019.
211. Wikipedia Contributors, "Citizen Science," Wikipedia, [Online]. Available: https://en.wikipedia.org/wiki/Citizen_science. [Accessed 31 December 2019].
212. R. Bonney, J. L. Shirk, T. B. Phillips, A. Wiggins, H. L. Ballard, A. J. Miller-Rushing and L. K. Parrish, "Next Steps for Citizen Science," *Science* , vol. 343, no. 6178, pp. 1436-1437, 2014.
213. L. Liebenberg, The Origin of Science: The Evolutionary Roots of Scientific Reasoning and its Implications for Tracking Science, Cape Town: CyberTracker, 2013/2021.
214. M. Santolini, "Covid-19: The Rise of a Global Collective Intelligence?," The Conversation, 24 April 2020. [Online]. Available: https://theconversation.com/covid-19-the-rise-of-a-global-collective-intelligence-135738. [Accessed 12 June 2020].
215. iNaturalist, "About," *iNaturalist* [Online]. Available: https://www.inaturalist.org. [Accessed 11 September 2023].
216. Wikipedia Contributors, "ECOSIA," Wikipedia, [Online]. Available: https://en.wikipedia.org/wiki/Ecosia [Accessed 19 February 2020].
217. M. N. Young and R. Leemans, "Group Report: Future Scenarios of Human-Environment Systems," in *Sustainability or Collapse? An Integrated History and Future of People on Earth*, Cambridge, The MIT Press, 2007, pp. 447-470.
218. MIT Management Sloan School, "The Climate Action Simulation: A Climate and Energy Policy Negotiation Game," [Online]. Available: https://mitsloan.mit.edu/teaching-resources-library/climate-action-simulation-a-climate-and-energy-policy-negotiation-game. [Accessed 26 May 2022].
219. T. L. Saaty, "The Three Laws of Thought, Plus One: The Law of Comparisons," *Axioms,* vol. 3, no. 1, pp. 46-49, 2014.
220. Wikipedia Contributors, "Syllogism," Wikipedia, [Online]. Available: https://en.wikipedia.org/wiki/Syllogism. [Accessed 19 September 2023].
221. D. Hitchcock, "Critical Thinking," The Stanford Encyclopedia of Philosophy, Edward N. Zalta (ed.), 2020. [Online]. Available: https://plato.stanford.edu/archives/fall2020/entries/critical-thinking/ Accessed 10-8-22. [Accessed 9 August 2022].
222. P. Shojaee, I. Mirzadeh, K. Alizadeh, M. Horton, S. Bengio and M. Farajtabar, "The Illusion of Thinking: Understanding the Strengths and Limitations of Reasoning Models Via the Lens of Problem Complexity," *arXiv preprint,* vol. arXiv:2506.06941, 2025.
223. Glenn and C. Jerome, "Participatory Methods, Futures Research Methodology V3.0, CD ROM/download," The Millennium Project, American Council for the United Nations University, 2009.

224. T. Gordon and J., "The Delphi Method, Futures Research Methodology V3.0, CD ROM/download," The Millennium Project, American Council for the United Nations University, 2009.
225. J. A. Morente-Molinera, G. Kou, K. Samuylov, F. J. Cabrerizo and E. Herrera-Viedma, "Using Argumentation in Expert's Debate to Analyse Multi-Criteria Group Decision Making Method Results," *Information Sciences,* vol. 573, pp. 433-452, 2021.
226. R. Biggs, A. De Vos, R. Preiser, H. Clements, K. Maciejewski and M. Schlüter, "Futures Analysis," in *The Routledge Handbook of Research Methods for Social-Ecological Systems*, Taylor & Francis, 2021.
227. Wikipedia Contributors, "Truth Table," Wikipedia, [Online]. Available: https://en.wikipedia.org/wiki/Truth_table. [Accessed 26 November 2022].
228. E. X. Li, "A Tale of Two Political Systems," TED Conferences, 2013. [Online]. Available: https://www.ted.com/talks/eric_x_li_a_tale_of_two_political_systems/. [Accessed 20 July 2022].
229. Wikipedia Contributors, "False Dilemma," Wikipedia, [Online]. Available: https://en.wikipedia.org/w/index.php?title=False_dilemma&oldid=1180240923. [Accessed 2 November 2023].
230. Y. Huang, "Why Democracy Still Wins: A Critique of Eric X. Li's "A Tale of Two Political Systems"," TED Blog, 1 July 2013. [Online]. Available: https://blog.ted.com/why-democracy-still-wins-a-critique-of-eric-x-lis-a-tale-of-two-political-systems/. [Accessed 26 November 2022].
231. M. Lapata, "What is Generative AI and How Does it Work?," The Royal Institution, London, 2023.
232. Veritasium, "The 4 Things it Takes to be an Expert," YouTube, 2 August 2022. [Online]. Available: https://www.youtube.com/watch?v=5eW6Eagr9XA. [Accessed 18 August 2022].
233. D. A. Schön, The Reflective Practitioner: How Professionals Think in Action, New York: Basic Books, 1983.
234. C. R. Sunstein and R. Hastie, Wiser: Getting Beyond Groupthink to Make Groups Smarter, Boston: Harvard Business Review Press, 2015.
235. D. A. Schon, Educating the Reflective Practitioner, San Francisco: Jossey-Bass, 1987.
236. J. Liu, H. Mooney, V. Hull, S. J. Davis, J. Gaskell, T. Hertel, J. Lubchenco and et. al., "Systems Integration For Global Sustainability," *Science ,* vol. 347, no. 6225 , p. 1258832, 2015.
237. A. Cassels, The Cochrane Collaboration: Medicine's Best-Kept Secret, Victoria: Agio Publishing House, 2015.
238. H. Pearson, "What's the Best Way to Tackle Climate Change? An 'Evidence Bank' Could Help Scientists Find Answers," Nature News, 6 June 2024. [Online]. Available: https://www.nature.com/articles/d41586-024-01683-4. [Accessed 10 June 2024].
239. R. G. Tiller, G. Destouni, M. Golumbeanu, Z. Kalantari, E. Kastanidi, L. Lazar, J.-M. Lescot and et. al., "Understanding Stakeholder Synergies Through System Dynamics: Integrating Multi-sectoral Stakeholder Narratives Into Quantitative Environmental Models," *Frontiers in Sustainability,* vol. 2, no. 701180, pp. 1-12, 2021.
240. D. Snowden, "Complex Acts of Knowing: Paradox and Descriptive Self-Awareness," *Journal of Knowledge Management,* vol. 6, no. 2, pp. 100-111, 2002.
241. Wikipedia Contributors, "Dunbar's Number," Wikipedia, [Online]. Available: https://en.wikipedia.org/wiki/Dunbar%27s_number. [Accessed 18 January 2021].
242. H. J. Miller, "Tobler's First law and Spatial Analysis," *Annals of the Association of American Geographers ,* vol. 94, no. 2, pp. 284-289, 2004.
243. A. Avena-Koenigsberger, J. Goni, R. Solé and O. Sporns, "Network Morphospace," *Journal of the Royal Society Interface,* vol. 12, no. 103, p. 20140881, 2015.
244. U. Özesmi and S. L. Özesmi, "Ecological Models Based on People's Knowledge: A Multi-step Fuzzy Cognitive Mapping Approach," *Ecological Modelling,* vol. 176, pp. 43-64, 2004.
245. S. Kininmonth, S. Gray and K. Kok, "Expert Modelling," in *The Routledge Handbook of Research Methods for Social-Ecological Systems*, New York, Routledge, 2022.

246. C. Becker, R. Chitchyan, L. Duboc, S. Easterbrook, M. Mahaux, B. Penzenstadler, G. Rodriguez-Navas and et. al., "The Karlskrona Manifesto for Sustainability Design," *arXiv preprint arXiv:1410.6968,* vol. Preprint, 2014.
247. A. Grubler, C. Wilson, N. Bento, B. Boza-Kiss, V. Krey, D. L. McCollum, N. D. Rao and et. al., "A Low Energy Demand Scenario for Meeting the 1.5 C Target and Sustainable Development Goals Without Negative Emission Technologies," *Nature Energy,* vol. 3, no. 6, pp. 515-527, 2018.
248. C. Alexander, The Nature of Order: An Essay on the Art of Building and the Nature of the Universe - Book 1: The Phenomenon of Life, Berkeley: The Center for Environmental Structure, 2002.
249. D. Bohm, "Five Ideas from David Bohm," Open Horizons, [Online]. Available: https://www.openhorizons.org/wholeness-and-the-implicate-order-five-ideas-from-david-bohm.html. [Accessed 22 August 2023].
250. S. Kauffman, At Home in the Universe: The Search for the Laws of Self-Organization and Complexity, Oxford: Oxford University Press, 1995.
251. M. Seligman, "Preface, Afterword," in *Homo Prospectus*, Oxford, Oxford University Press, 2016, pp. 89, 352.
252. S. French and N. Argyris, "Decision Analysis and Political Processes," *Decision Analysis ,* vol. 15, no. 4, pp. 208-222, 2018.
253. L. Kemp, C. Xu, J. Depledge, K.L. Ebi, G. Gibbins, T.A. Kohler, J. Rockström, M. Scheffer, H.J. Schellnhuber, W. Steffen and T.M. Lenton, "Climate Endgame: Exploring Catastrophic Climate Change Scenarios," *Proceedings of the National Academy of Sciences,* vol. 119, no. 34, p. e2108146119, 2022.
254. Wikipedia Contributors, "Greenhouse Gas Inventory," Wikipedia, [Online]. Available: https://en.wikipedia.org/wiki/Greenhouse_gas_inventory. [Accessed 9 September 2023].
255. M. Lewis, The Fifth Risk: Undoing Democracy, London: Allen Lane , 2018.
256. B. J. de Vries, "Scenarios: Guidance for an Uncertain and Complex World?," in *Sustainability or Collapse? An Integrated History and Future of People on Earth*, Cambridge, MA, The MIT Press, 2007, pp. 380-397.
257. G. C. Gallopín, "Planning for Resilience: Scenarios, Surprises, and Branch Points," in *Panarchy: Understanding Transformations in Human and Natural Systems*, Washington, Island Press, 2002, pp. 361-392.
258. M. Mahmoud, Y. Liu, H. Hartmann, S. Stewart, T. Wagener, D. Semmens, R. Stewart, H. Gupta, D. Dominguez, F. Dominguez, D. Hulse, R. Letcher, B. Rashleigh, C. Smith and R. Street, "Formal Framework for Scenario Development in Support of Environmental Decision-Making," *Environmental Modelling and Software,* vol. 24, pp. 798-808, 2009.
259. H. Lübker, "Envisioning Alternative Futures Helps Rethink Human-Ocean Relationships," Stockholm Resilience Center, 8 June 2023. [Online]. Available: https://www.stockholmresilience.org/research/research-news/2023-06-08-envisioning-alternative-futures-helps-rethink-human-ocean-relationships.html. [Accessed 18 June 2023].
260. C. Sunter and M. Ilbury, Thinking the Future: New Perspectives from the Shoulders of Giants, Cape Town: Penguin Books, 2021.
261. D. Yergin, The New Map: Energy, Climate, and the Clash of Nations, London: Allen Lane, 2020.
262. H.O. Pörtner, D.C. Roberts, H. Adams, I. Adelekan, C. Adler, R. Adrian, P. Aldunce, E. Ali, R. Ara Begum, B. Bednar-Friedl and R. Bezner Kerr, "Climate Change 2022: Impacts, Adaptation, and Vulnerability," Cambridge University Press, Cambridge, 2022.
263. HM Treasury, "The Economics of Biodiversity: The Dasgupta Review, Headline Messages," UK Government, London, 2021.
264. J. E. Stiglitz, The Great Divide, London: Penguin Books, 2015.
265. Wikipedia Contributors, "Fourth Industrial Revolution," Wikipedia, [Online]. Available: https://en.wikipedia.org/wiki/Fourth_Industrial_Revolution. [Accessed December 29 2022].
266. International Energy Agency, "Clean Energy Innovation," IEA, Paris, 2020.

267. A.J. Constable, S., French, V. Karoblyte and D. Viner, "Decision-Making for Managing Climate-Related Risks: Unpacking the Decision Process to Avoid "Trial-and-Error" Responses," *Frontiers in Climate,* vol. 4, p. 754264, 2022.
268. Intergovernmental Panel on Climate Change, "AR6 Synthesis Report: Summary for Policymakers Headline Statements," IPCC, 2023. [Online]. Available: https://www.ipcc.ch/report/ar6/syr/resources/spm-headline-statements/. [Accessed 30 October 2023].
269. H. Rosling, O. Rosling and A. Rösling Ronnlund, Factfulness: Ten Reasons We're Wrong About the World – And Why Things Are Better Than You Think, London: Sceptre, 2018.
270. T. Piketty, Capital in the Twenty-First Century, Cambridge: The Belknap Press, 2017.
271. A. Gore, Presenter, *What the Fossil Fuel Industry Doesn't Want You to Know.* [Performance]. TED Countdown Summit, 2023.
272. C. McGlade and P. Ekins, "The Geographical Distribution of Fossil Fuels Unused When Limiting Global Warming to 2 C," *Nature,* vol. 517, no. 7533, pp. 187-190, 2015.
273. International Energy Agency, "Special Report on Clean Energy Innovation: Accelerating Technology Progress for a Sustainable Future," IEA Publications, Paris, 2020.
274. K. L. Tang A, "A Fate Worse Than Warming? Stratospheric Aerosol Injection and Global Catastrophic Risk," *Frontiers in Climate Science,* vol. 3, pp. 1-17, 2021.
275. S. Whitehouse, "Senator Whitehouse Makes the Case Against Kash Patel for FBI Director," YouTube, 20 February 2025. [Online]. Available: https://www.youtube.com/watch?v=F_MNAT9DgLk. [Accessed 21 February 2025].
276. United Nations Environment Programme, "Net-Zero Banking Alliance," UNEP Finance Initiative, [Online]. Available: https://www.unepfi.org/net-zero-banking/members/. [Accessed 21 February 2025].
277. E. Elhacham, L. Ben-Uri, J. Grozovski, Y. M. Bar-On and R. Milo, "Global Human-made Mass Exceeds All Living Biomass," *Nature* , vol. 588, no. 7838, pp. 442-444., 2020.
278. J.-F. Bastin, Y. Finegold, C. Garcia, D. Mollicone, M. Rezende, D. Routh, C. M. Zohner and T. W. Crowther, "The Global Tree Restoration Potential," *Science,* vol. 365, no. 6448, pp. 76-79, 2019.
279. World Economic Forum, "Top 10 Emerging Technologies of 2023, Flagship Report," World Economic Forum, Geneva, 2023.
280. T. W. Murphy Jr, "Limits to Economic Growth," *Nature Physics,* vol. 18, no. 8, pp. 844-847, 2022.
281. M. Barnard, "Why Aren't Energy Flows Diagrams Used More to Inform Decarbonization?," CleanTechnica, 13 February 2023. [Online]. Available: https://cleantechnica.com/2023/02/13/why-arent-energy-flows-diagrams-used-more-to-inform-decarbonization/. [Accessed 15 February 2023].
282. J. Hickel, P. Brockway, G. Kallis, L. Keyßer, M. Lenzen, A. Slameršak, J. Steinberger, Ürge-Vorsatz and Diana, "Urgent Need for Post-Growth Climate Mitigation Scenarios," *Nature Energy,* vol. 6, no. August, pp. 766-768, 2021.
283. R. Costanza, R. d'Arge, R. de Groot, S. Farberk, M. Grasso, B. Hannon, K. Limburg, S. Naeem, R. V. O'Neill, J. Paruelo, R. G. Raskin, P. Suttonkk and M. van den Belt, "The Value of the World's Ecosystem Services and Natural Capital," *Nature,* vol. 387, no. 6630, pp. 253-260, 1997.
284. J. E. Stiglitz, A. Sen and J.-P. Fitoussi, "Report by the Commission on the Measurement of Economic Performance and Social Progress," 2009.
285. T. Alexander, M.C.H. Dziobek and T. Galeza, Sustainable Development Goals (SDGs) and GDP: What National Accounts Bring to the Table, International Monetary Fund, 2018.
286. D. Estlund, "Democracy Counts," in *Collective Wisdom: Principles and Mechanisms*, New York, Cambridge University Press, 2014, pp. 230-250.
287. R. E. Goodin, "Enfranchising All affected Interests, and Its Alternatives," *Philosophy and Public Affairs,* vol. 35, no. 1, pp. 40-68, 2007.
288. J. M. Polimeni, K. Mayumi, M. Giampietro and B. Alcott, The Myth of Resource Efficiency: The Jevons Paradox, New York: Earthscan, 2009.

289. L. Mari, "Epistemology of Measurement," *Measurement,* vol. 34, no. 1, pp. 17-30, 2003.
290. J. Ball, "The Double Diamond: A Universally Accepted Depiction of the Design Process," Design Council, [Online]. Available: https://www.designcouncil.org.uk/news-opinion/double-diamond-universally-accepted-depiction-design-process. [Accessed 3 December 2019].
291. M. Van Wyk and E. Gideon, "The Relevance of Integrated Reporting in Future Standard Setting of the International Sustainability Standards Board," *Frontiers in Sustainability,* no. 4: 1218985, 2023.
292. E. T. Hall, Beyond Culture, Garden City: Anchor Press/Doubleday & Company, 1976.
293. D. Meder, F. Rabe, T. Morville, K. H. Madsen, M. T. Koudahl, R. J. Dolan, H. R. Siebner and O. J. Hulme, " Ergodicity-breaking Reveals Time Optimal Decision Making in Humans," *PLOS Computational Biology* , vol. 17, no. 9, p. e1009217, 2021.
294. S. Wolfram, A New Kind of Science, Champaign: Wolfram Media, 2001.
295. M. G. Moore, "How to Make Great Decisions, Quickly," Harvard Business Review, 22 March 2022. [Online]. Available: https://hbr.org/2022/03/how-to-make-great-decisions-quickly. [Accessed 2 November 2023].
296. R. Baumeister, "Pragmatic Prospection," in *Homo Erectus*, New York, Oxford University Press, 2016.
297. E. O. Wilson, Half-Earth: Our Planet's Fight for Life, New York: Liveright Publishing Corporation, 2016.
298. K. De Meyer, E. Coren, M. McCaffrey and C. Slean, "Transforming the Stories We Tell About Climate Change: From 'Issue' to 'Action'," *Environmental Research Letters,* vol. 16, no. 1, p. 015002, 2020.
299. F. Creutzig, J. Roy, P. Devine-Wright, J. Díaz-José, F. Geels, A. Grubler, N. Maïz, E. Masanet, Y. Mulugetta, C. Onyige, P. Perkins, A. Sanches-Pereira and E. Weber, "Climate Change 2022: Mitigation of Climate Change. Contribution of Working Group III to the Sixth Assessment Report of the Intergovernmental Panel on Climate Change," Cambridge University Press, Cambridge, 2022.
300. M. Rees, On the Future Prospects for Humanity, Princeton: Princeton University Press, 2018, p. 213.
301. United Nations Educational, Scientific and Cultural Organization, "Philosophy: A School of Freedom," UNESCO, Paris, 2007.
302. F. A. Hassan, "The Lie of History: Nation-States and the Contradictions of Complex Societies," in *Sustainability or Collapse: An Integrated History and Future of People on Earth*, London, The MIT Press, 2007, pp. 169-196.
303. P. Barkham, "Should Rivers Have the Same Rights as People?," The Guardian, 25 July 2021. [Online]. Available: https://www.theguardian.com/environment/2021/jul/25/rivers-around-the-world-rivers-are-gaining-the-same-legal-rights-as-people. [Accessed 5 October 2023].
304. United Nations, "UN General Assembly Declares Access to Clean and Healthy Environment a Universal Human Right," 28 07 2022. [Online]. Available: https://news.un.org/en/story/2022/07/1123482. [Accessed 28 August 2025].
305. Lex Fridman Podcast, "Accessed 18/3/2025 Narendra Modi: Prime Minister of India - Power, Democracy, War & Peace," YouTube, March 2025. [Online]. Available: https://www.youtube.com/watch?v=ZPUtA3W-7_I&t=11281s. [Accessed 18 March 2025].

Index

E. Bellos, *Sustainability Thinking*, https://doi.org/10.1007/978-3-032-11244-6

MIX
Papier aus verantwortungsvollen Quellen
Paper from responsible sources
FSC® C105338

If you have any concerns about our products,
you can contact us on
ProductSafety@springernature.com

In case Publisher is established outside the EU,
the EU authorized representative is:
Springer Nature Customer Service Center GmbH
Europaplatz 3, 69115 Heidelberg, Germany

Printed by Libri Plureos GmbH
in Hamburg, Germany